Lecture Notes in Computational Science and Engineering

22

Springer
Berlin
Heidelberg
New York
Barcelona
Hong Kong
London
Milan
Paris
Tokyo

Karsten Urban

Wavelets in Numerical Simulation

Problem Adapted Construction and Applications

With 61 Figures

Springer

Karsten Urban
RWTH Aachen
Institut für Geometrie
und Praktische Mathematik
Templergraben 55
52056 Aachen, Germany
e-mail: urban@igpm.rwth-aachen.de

Cataloging-in-Publication Data applied for

Die Deutsche Bibliothek - CIP-Einheitsaufnahme

Urban, Karsten:
Wavelets in numerical simulation : problem adapted construction
and applications / Karsten Urban. - Berlin ; Heidelberg ; New York ;
Barcelona ; Hong Kong ; London ; Milan ; Paris ; Tokyo : Springer, 2002
(Lecture notes in computational science and engineering ; Vol. 22)
ISBN 3-540-43055-5

Mathematics Subject Classification (2000):
65T60, 65N30, 65N55, 35J05, 35J20, 35J60, 35Q30, 35Q60, 35Q72

ISSN 1439-7358
ISBN 3-540-43055-5 Springer-Verlag Berlin Heidelberg New York

Springer-Verlag Berlin Heidelberg New York
a member of BertelsmannSpringer Science + Business Media GmbH
http://www.springer.de

Printed in Germany

Cover Design: Friedhelm Steinen-Broo, Estudio Calamar, Spain
Cover production: *design & production*
Typeset by the author using a Springer TEX macro package

Printed on acid-free paper SPIN: 10859435 46/3142/LK - 5 4 3 2 1 0

To Almut, Tobias, Niklas and Annika.

Preface

Sapere aude!

Immanuel Kant (1724-1804)

Numerical simulations play a key role in many areas of modern science and technology. They are necessary in particular when experiments for the underlying problem are too dangerous, too expensive or not even possible. The latter situation appears for example when relevant length scales are below the observation level. Moreover, numerical simulations are needed to control complex processes and systems. In all these cases the relevant problems may become highly complex. Hence the following issues are of vital importance for a numerical simulation:

- Efficiency of the numerical solvers:
  Efficient and fast numerical schemes are the basis for a simulation of 'real world' problems. This becomes even more important for realtime problems where the runtime of the numerical simulation has to be of the order of the time span required by the simulated process. Without efficient solution methods the simulation of many problems is not feasible. 'Efficient' means here that the overall cost of the numerical scheme remains proportional to the degrees of freedom, i.e., the numerical approximation is determined in *linear time* when the problem size grows e.g. to upgrade accuracy. Of course, as soon as the solution of large systems of equations is involved this requirement is very demanding.
- Accuracy and reliability of the numerical approximation:
  Even though in many cases it is possible to validate numerical simulations by comparisons with experimental data, these methods will be applied to problems where no knowledge of the behavior of the solution is available. In particular when design processes are based upon numerical simulations, the method has to produce a reliable and accurate approximation.
- Data compression:
  Given efficient numerical schemes, the complexity of 'real world' problems

from science and engineering is often still so large that 'ad hoc' discretization methods would overflow memory capacities even of computers of the next generations. Without efficient strategies for keeping the discrete problems as small as possible without sacrificing too much accuracy, these large scale and sometimes high-dimensional problems can not be treated appropriately. Of course, a sound mathematical analysis is required to determine the best accuracy/work balance that can be realized for a given problem.

All these issues are partly correlated and closely intertwined with the mathematical tools a numerical method is based upon. It is desirable that a tool allows the combination of these issues. In particular, the last issue in the above wish lists calls for the design of schemes that automatically adapt to a given problem at hand, so that the desired overall accuracy of the approximate solution is obtained at the expense of possibly few degrees of freedom in the discretization. For instance, in the context of finite elements one would want to employ a small mesh size where ever the approximated object exhibits small scale features or singularities. Such concepts have been studied and developed in various discretization settings such as finite volume, finite difference or finite element methods. Although by no means common yet in industrial applications these methods have achieved already a certain level of practical maturity that strongly encourages further efforts in this direction. On the other hand, on the level of a rigorous analysis many problems appear to remain widely open. Given such adaptive concepts, again it remains to efficiently solve the resulting hopefully reduced systems of equations which however may now exhibit somewhat different features than their counterparts based on uniform refinements. The multigrid method is a prominent example of a general methodology to deal with this task by exploiting the interaction of different scales. In many cases this works very well although matters are no longer covered by the available analysis.

Wavelets offer an alternative that perhaps to a somewhat larger extent offers a synthesis of the discretization and solution process in an adaptive solution framework. This is mainly due to what we will sometimes refer to as 'strong analytical properties' of wavelet bases mainly based on their good *localization* in frequency and physical space, their *cancellation properties* that entail quasi sparse representations of many relevant operators and of functions with isolated singularities, and last but not least on the fact that wavelet expansions induce isomorphisms between many function spaces and sequence spaces which offers a natural coupling between the continuous and discrete realm. Based on these properties a rigorous convergence and cost analysis for adaptive wavelet methods has been recently developed, [28, 29, 39, 42, 43]. In particular, it was proven in [28, 29] that for a large class of operator equations an appropriate adaptive wavelet method converges at an optimal rate with almost optimal complexity.

The prize to be paid for these features is the higher sophistication of the tool itself. This accounts to some extent for the fact that the application of these concepts to real life problems is still much less advanced than for the above mentioned more conventional schemes. In fact, the applicability of a wavelet basis for the numerical simulation of a given problem demands, of course, the availability of an appropriate basis. This requirement is twofold: First the bases should be adapted to the considered *problem* and second to the *domain* of interest. Adapting a basis to a problem is not only done by choosing functions with the appropriate smoothness. Many problems pose extra conditions that have to be met by a discretization. In this monograph we focus on both aspects, namely on the construction of appropriate wavelet bases even on complex domains and on the adaptation of such wavelet bases to differential operators involving the div- and **curl**-operator.

We are now going to describe the organization of this monograph in somewhat more detail:

In Chap. 1 we describe a construction of wavelet bases also on fairly general domains. This construction is based upon a domain decomposition and mapping strategy. The domain of interest is subdivided into non-overlapping subdomains which are mapped to a single reference cube. On the cube, tensor products of appropriate wavelet systems on the interval $(0, 1)$ are used. This is described in Sect. 1.3. However, the understanding of the subsequent developments does not require working through all details of the construction. The important point is that these systems of functions are built in such a manner that certain relevant properties hold. We collect in Sect. 1.1 those properties that will later be needed for our analysis. Those who are not interested in the technical details of the construction are therefore invited to consult just this section before moving on to the subsequent topics. It should be mentioned that most of the described constructions are already realized in public domain software packages.

Chap. 2 is devoted to the adaptation of wavelet bases to differential problems in $\boldsymbol{H}(\mathrm{div};\Omega)$ and $\boldsymbol{H}(\mathbf{curl};\Omega)$. In fact, another potential advantage is the flexibility of biorthogonal wavelet bases. As opposed to *orthonormal* wavelet bases the concept of *biorthogonal* systems leaves some freedom in the construction while still ensuring all relevant properties. This freedom can be used to fulfill additional requirements that are needed for the discretization of problems in $\boldsymbol{H}(\mathrm{div};\Omega)$ and $\boldsymbol{H}(\mathbf{curl};\Omega)$. These spaces arise naturally in the variational formulation of a whole variety of partial differential equations. We construct biorthogonal wavelet bases for these spaces. In particular, we construct divergence- and curl-free wavelet bases which give rise to an orthogonal Hodge decomposition of spaces of vector fields. It should be noted that the properties of this Hodge decomposition are inherited by any finite dimensional subspaces spanned by any subsets of the full wavelet bases. This implies that for each (adaptive) selection of wavelet functions certain operators can be decoupled. Moreover, divergence- and curl-free wavelets can directly be used to discretize problems involving this as an extra condition. For instance, with regard to the incompressible Navier-Stokes equations the use of divergence-free wavelets allows one (at least on 'simple' domains) to replace the somewhat more complex saddle point problem by an elliptic one. By a fictitious domain approach these bases can also be used on complex domains – at the expense of having to deal again with a saddle point problem induced by the Lagrange multipliers for the boundary conditions. However, the dimension of the Lagrange multiplier spaces is in this case much smaller than that for the trial spaces on the domain.

In Chap. 3 we describe various applications of our constructions. We consider examples such as the Lamé equations for nearly incompressible material in linear elasticity, the incompressible Navier-Stokes equations in fluid dynamics and Maxwell's equations in electromagnetism. In all these equations, certain parameters appear that may come from the problem itself or from a discretization in time. First, we use the analytical properties of the wavelet bases used for the discretization in order to prove *robustness* and *optimality* of certain wavelet preconditioners. By *robust* we mean, that the condition of the preconditioned system should be independent of the involved parameters. By *optimal* we mean that the condition number of the arising linear systems is independent of the number of unknowns. This follows from the fact that (a properly scaled) wavelet representation of the differential operator defines a boundedly invertible mapping on ℓ_2 combined with the stability of Galerkin discretizations (which is for granted in connection with definite problems). Realizing these properties for the above function spaces is an important prerequisite for the application of the results in [28, 29] in this context.
While these applications are of more theoretical nature, the next two are of more experimental character. In Sect. 3.2 we use divergence-free wavelets for the analysis and simulation of turbulent flows. We show that the analytical properties of these wavelets offer a powerful tool both for the simulation of incompressible flows as well as for the analysis of given flow data. Using these bases allows one to resolve local effects and to separate errors in the data due to inexact measuring or approximate modeling. The multiscale structure of such flow fields is shown. Finally, we investigate the smoothness of the given flow fields in a certain *Besov* scale which is an indicator for the potential complexity reduction offered by an adaptive Wavelet-Galerkin method.

Finally, in Sect. 3.3 we study the hardening of an elastoplastic rod. The occurance of plasticity is due to certain local conditions on the stress of the material. These conditions involve point values of the stress. On the other hand, the displacement of the elastic part is determined by an elliptic partial differential equation. In order to combine the advantages of wavelet discretizations of elliptic problems with the pointwise correction, we use an elastic predictor-plastic corrector method with interpolatory wavelets for the stress correction. In this framework, we use the analytical properties of wavelet bases in order to detect plastic waves, i.e., regions of plasticity, directly from the wavelet coefficients. It turns out that a certain discrete sequence of weightened wavelet coefficients allows us to clearly identify plastic regions.

Acknowledgments: It is impossible to list all friends and colleagues that supported me and my work over the last years leading also to this monograph. I have to restrict myself to some few of them knowing and regretting the fact that many others also deserve mentioning.

I am very grateful to Wolfgang Dahmen for the supervision of my work over the last nine years. Without all the inspiring discussions and several hints and advises, this monograph would not be as it is. I also thank the Professors Henning Esser (Aachen), Reinhold Schneider (Chemnitz) and Claudio Canuto (Turin) for acting as referees for my habilitation thesis at the RWTH Aachen which is the basis of this monograph.

Not only my scientific career has been highly influenced by my friends and colleagues in Italy, in particular from Torino, Pavia and Milano. I am grateful for the kind hospitality and the cooperation before, during and after my permanent stay in Italy. In particular, I want to mention my friend Claudio Canuto. I am extremely grateful for the cooperation with him and all his support over several years.

I would like to thank my colleagues at the Institut für Geometrie und Praktische Mathematik, RWTH Aachen, the Istituto di Analisi Numerica del C.N.R., Pavia and at the Mathematisch Instituut of the Universiteit Utrecht for the nice atmosphere in all these institutes and for so many occasions where I received help.

Work is important, but it is not all. My family has always been the center of my life. My parents, my wife Almut and my children Tobias, Niklas and Annika are giving me the energy also for my scientific work.

Aachen, December 2001 *Karsten Urban*

Table of Contents

1 Wavelet Bases

In this chapter, we describe the general framework of wavelet bases and review some of their particular examples. We will first give the definition and some important features in a rather general setting. This will be useful since this general framework fits to the various applications we will consider later. Then we describe wavelet bases starting on the real line up to general domains.

Originally, wavelets have first been constructed on the real line or —in higher dimensions— on the whole Euclidean space. These constructions are of course only of limited use when one aims at numerically solving an operator equation on a bounded domain or manifold. The construction of wavelet bases for general bounded domains and manifolds has been an area of very active research during the past years so that nowadays wavelet bases are available on fairly general domains for a wide class of applications. We will describe some of these constructions here.

After describing the abstract setting, we will start by the construction of wavelet bases on the real line in Sect. 1.2. Based on this we proceed to the unit interval $(0,1)$ in Sect. 1.3. Already in this situation one can clearly identify the main difficulties when going from the Euclidean space to a bounded domain. Moreover, we will use wavelet bases on $(0,1)$ in Sect. 1.4 in a tensor product approach to obtain wavelets for the n-dimensional cube. This in turns will be used in Sect. 1.5 to construct wavelet bases on fairly general domains by using domain decomposition and mapping ideas. Vector valued wavelets will be described in Sect. 1.6.

If the reader is not interested in all details of the construction, it suffices to read Sect. 1.1 in order to proceed to the various applications. In this section, we collect all relevant properties.

Here and in the sequel, we will often use the notation $A \lesssim B$ to abbreviate $A \leq cB$ with some constant $c > 0$, and for $A \lesssim B \lesssim A$, we write $A \sim B$.

1.1 Wavelet Bases in $L_2(\Omega)$

In this section, we collect from [44, 45, 46] all relevant properties and the main results of wavelet bases in $L_2(\Omega)$ and in subspaces of $L_2(\Omega)$. In Sect. 1.1.1, we describe the general setting of wavelet bases in subspaces of $L_2(\Omega)$ and show a useful tool for their construction, namely multiresolution. Under certain approximation and regularity assumptions on the multiresolution spaces a characterization of Sobolev spaces in terms of projectors associated with the multiresolution can be shown, which is reviewed in Sect. 1.1.2. In order to show the Riesz basis property of wavelet systems in $L_2(\Omega)$ in Sect. 1.1.3, biorthogonality turns out to be a necessary condition. We review wavelet characterizations of Sobolev spaces and collect further features of wavelet bases that will be used in the sequel.

1.1.1 General Setting

Let us consider a Banach subspace $H \hookrightarrow L_2(\Omega)$ normed by $\|\cdot\|_H$. In this section we review the general setting of wavelet bases for H. Particular examples of H will be Sobolev- and Besov spaces as well as the spaces $\boldsymbol{H}(\mathrm{div};\Omega)$, $\boldsymbol{H}(\mathbf{curl};\Omega)$ and closed subspaces of those for domains $\Omega \subset \mathbb{R}^n$. The subsequent discussions follow similar lines in [44, 45, 46] where a general program is initiated for identifying on one hand relevant features of wavelet bases for applications to operator equations and on the other hand for deriving criteria for the realization of such features in corresponding settings, in particular, to obtain substitutes for classical Fourier techniques that are usually used in the more classical shift-invariant setting on $\mathbb{R}^n$ or the torus $\mathbb{T}^n = \mathbb{R}^n/\mathbb{Z}^n$.

Before we give the definition of a wavelet basis, let us describe the character of these functions which is also reflected by the way they will be labeled. We consider a family of functions

$$\Psi := \{\psi_\lambda : \lambda \in \mathcal{J}\}, \tag{1.1}$$

where $\mathcal{J}$ denotes some infinite set of indices. Each index $\lambda \in \mathcal{J}$ takes the form $\lambda = (j,k)$, where

$$|\lambda| := j \in \mathbb{Z} \tag{1.2}$$

denotes the *scale* or *level* and k represents the location of ψ_λ in space as well as the type of element (see below). Moreover, we assume there exist infinitely many non-empty subsets $\mathcal{J}_j := \{\lambda \in \mathcal{J} : |\lambda| = j\}$ so that Ψ is a *multiscale* system.

Definition 1. *A family Ψ as in* (1.1) *is called a* generalized wavelet basis *for H, if*

a.) Ψ is generalized Riesz basis, *i.e., it spans H*

$$H = \mathrm{clos}_{\|\cdot\|_H} \operatorname{span} \Psi \tag{1.3}$$

and there exists an invertible operator $\boldsymbol{D} \in [\ell_2(\mathcal{J}), \ell_2(\mathcal{J})]$ such that

$$\|\boldsymbol{d}^T \Psi\|_H := \Big\| \sum_{\lambda \in \mathcal{J}} d_\lambda \psi_\lambda \Big\|_H \sim \|\boldsymbol{D}\,\boldsymbol{d}\|_{\ell_2(\mathcal{J})} = \left(\sum_{\lambda \in \mathcal{J}} |(\boldsymbol{D}\,\boldsymbol{d})_\lambda|^2 \right)^{1/2}, \tag{1.4}$$

for $\boldsymbol{d} := \{d_\lambda\}_{\lambda \in \mathcal{J}}$.

b.) The functions are local in the sense that the supports of ψ_λ scale like

$$\operatorname{diam}(\operatorname{supp}\psi_\lambda) \sim 2^{-|\lambda|}, \quad \lambda \in \mathcal{J}. \tag{1.5}$$

If the operator $\boldsymbol{D}$ in (1.4) *is diagonal, we call Ψ a* wavelet basis *for H.* □

Some remarks concerning the above definition are in order.

- Note that as opposed to other definitions of wavelet bases we only consider *compactly supported* functions here. This is due to the fact that we ultimately aim at using wavelets as trial and test functions in a Galerkin method for numerically solving certain operator equations.
- The second remark concerns the generalized Riesz basis property in (1.4). For $H = L_2(\Omega)$ and $\boldsymbol{D} = \boldsymbol{I}$, (1.4) is just the usual Riesz basis estimate. However, we will also treat systems that fulfill (1.4) for a closed subspace H of $L_2(\Omega)$, but not for $L_2(\Omega)$ itself. This is the reason why we setup the definition in this way. On the other hand, we will also describe various examples where (1.4) is valid for a whole family of spaces H in particular including $L_2(\Omega)$.
- As we shall see in Theorem 2 on p. 9 below, in may cases $\boldsymbol{D}$ is a *diagonal* matrix. In this case, (1.4) implies that a properly scaled system forms a Riesz basis for H. This is an important point which is often used in applications, see, e.g., Theorem 13 on p. 111 below.
- If $\boldsymbol{D}$ is *not* diagonal, it represents a change of bases in the sense that (1.4) implies that $\boldsymbol{D}^{-T}\Psi$ is a Riesz basis for H. However, we decided to choose also this slightly more general definition here since it is well suited to the spaces $\boldsymbol{H}(\mathrm{div};\Omega)$, $\boldsymbol{H}(\mathbf{curl};\Omega)$ and the subspaces of divergence-, resp. curl-free vector fields in Chap. 2 below.

Scaling Systems and Multiresolution. In many cases, the wavelet system Ψ is constructed with the aid of a family of collections of functions

$$\Phi_j := \{\varphi_{j,k} : k \in \mathcal{I}_j\} \subset H,$$

where again $j \in \mathbb{Z}$ can be understood as the scale or the level and $\mathcal{I}_j$ is some appropriate set of indices encoding the spatial location of (the compactly supported) functions $\varphi_{j,k}$. Moreover, these collections will always be assumed to form *stable bases* of the closure of their span in the sense that

$$\|\{c_{j,k}\}_{k\in\mathcal{I}_j}\|_{\ell_2(\mathcal{I}_j)} \sim \left\| \sum_{k\in\mathcal{I}_j} c_{j,k}\,\varphi_{j,k} \right\|_{0,\Omega}, \tag{1.6}$$

where $\|\cdot\|_{0,\Omega}$ denotes the standard norm in $L_2(\Omega)$. Viewing Φ_j as a column vector with components $\varphi_{j,k}$, $k \in \mathcal{I}_j$, we say that $\{\Phi_j\}_{j\in\mathbb{N}_0}$ is *refinable* if there exists an operator $M_{j,0} \in [\ell_2(\mathcal{I}_j), \ell_2(\mathcal{I}_{j+1})]$ such that

$$\Phi_j = M_{j,0}^T \Phi_{j+1}, \tag{1.7}$$

[25]. Often, $M_{j,0}$ is also called *refinement matrix* (even though it might be an infinite matrix). In other words, (1.7) (which is also called *two-scale relation* or *refinement equation*) means that each function on level j can be represented as a linear combination of the basis functions on level $j+1$, i.e.,

$$\varphi_{j,k} = \sum_{k' \in \mathcal{I}_{j+1}} m^j_{k',k} \varphi_{j+1,k'}, \qquad M_{j,0} := (m^j_{k',k})_{k' \in \mathcal{I}_{j+1}, k \in \mathcal{I}_j}. \tag{1.8}$$

If the stability constants in (1.6) do not depend on j, a refinable system is also called *single scale system*, *generator* or *scaling* system. The functions $\varphi_{j,k}$ are called *single scale basis*, *scaling functions* or *generators*. The refinement equation (1.7) in particular implies that the induced spaces

$$S_j := S(\Phi_j) := \operatorname{clos}_{L_2(\Omega)} \operatorname{span}(\Phi_j) \tag{1.9}$$

are *nested*: $S_j \subset S_{j+1}$. As in the classical shift-invariant setting, the sequence $\mathcal{S} := \{S_j\}_{j\in\mathbb{Z}}$ of these spaces is often called *multiresolution analysis*, [44, 46, 103].

Before progressing, let us add some remarks on multiresolution analyses. As we shall see, a construction of wavelets is usually based upon such a sequence. Moreover, it is well-known that also certain hierarchies of finite elements or spline functions form multiresolution analyses.

Finally, a comment on the index sets $\mathcal{I}_j$ is in order. In most of our applications, $\mathcal{I}_j$ will be a finite set. Then, one does of course not need to consider the closure for example in the definition (1.9) of S_j.

Complement Spaces. Next, one defines *complement spaces* W_j by

$$W_j := S_{j+1} \ominus S_j, \tag{1.10}$$

where –so far– we do not fix in which sense the complement will have to be understood. We will discuss this issue later. The wavelet functions have to be chosen in such a way to form a basis for W_j, i.e.,

$$W_j = S(\Psi_j), \qquad \Psi_j := \{\psi_{j,k} : k \in \mathcal{J}_j\},$$

where again $\mathcal{J}_j$ is a suitable set of indices such that $\#\mathcal{J}_j = \#\mathcal{I}_{j+1} - \#\mathcal{I}_j$. Hence, W_j is also called *wavelet space*. In order to ensure that

$$\Psi := \bigcup_{j\geq -1} \Psi_j = \{\psi_\lambda : \lambda \in \mathcal{J}\}, \qquad \mathcal{J} := \bigcup_{j\geq -1} \bigcup_{k\in\mathcal{J}_j} (j,k),$$

($\Psi_{-1} := \Phi_0$, $\mathcal{J}_{-1} := \mathcal{I}_0$) forms a Riesz basis for H, each Ψ_j has to form a Riesz basis for W_j. However, this is not sufficient.

1.1.2 Characterization of Sobolev-Spaces

It has been shown in [44, 45, 46] that (aside from the necessary condition of biorthogonality, see Sect. 1.1.3 on p. 7 below) the Riesz basis property of a multiscale basis can be asserted by further conditions that are independent of these bases but refer to properties of the multiresolution spaces. These conditions can be phrased for classical Sobolev spaces in terms of Jackson and Bernstein estimates as will be explained next. For the role of boundary conditions in this context, see [58].

Jackson Inequality. For many examples of biorthogonal multiresolution sequences, one can show the following inequality for $v \in H^s(\Omega)$, [47]

$$\inf_{v_j \in S_j} \|v - v_j\|_{0,\Omega} \lesssim 2^{-sj} \|v\|_{s,\Omega}, \quad 0 \le s \le d_S. \tag{1.11}$$

Here, $\|\cdot\|_{s,\Omega}$ defines the norm in the Sobolev space $H^s(\Omega)$ defined in the following standard way (see e.g. [47]): For $m \in \mathbb{N}$, let

$$\|u\|^2_{m,\Omega} := \sum_{|\alpha| \le m} \|\partial^\alpha u\|^2_{0,\Omega},$$

where $v := \partial^\alpha u$ denotes the *weak derivative* of order α, i.e.,

$$(\phi, v)_{0,\Omega} = (-1)^{|\alpha|} (\partial^\alpha \phi, u)_{0,\Omega},$$

for all $\phi \in C_0^\infty(\Omega)$, where

$$(f, g)_{0,\Omega} := \int_\Omega f(x)\, g(x)\, dx, \qquad f, g \in L_2(\Omega),$$

see, e.g., [18]. For non-integer values $s \in \mathbb{R} \setminus \mathbb{N}$, one can define $H^s(\Omega)$ by interpolation [11, 69]. Alternatively, one can set

$$\|u\|_{s,\Omega} := \inf\{\|g\|_{s,\mathbb{R}^n} : g \in H^s(\mathbb{R}^n), g_{|\Omega} = u\}.$$

Here, the Sobolev norm on $\mathbb{R}^n$ is defined by

$$\|g\|^2_{s,\mathbb{R}^n} := \int_{\mathbb{R}^n} (1 + |y|^2)^s\, |\mathcal{F}g(y)|^2\, dy,$$

where

$$\mathcal{F}g(y) := \int_{\mathbb{R}^n} g(\xi)\, e^{-i\xi \cdot y}\, d\xi$$

is the Fourier transform and $H^s(\mathbb{R}^n) := \{f \in L_2(\mathbb{R}^n) : \|f\|_{s,\mathbb{R}^n} < \infty\}$.

Note that this inequality is a property of the multiresolution spaces S_j. The wavelet spaces play no role here. In the case of simple domains, one can often show (1.11) for $d_S = L$, where L denotes the order of exactness of the system, i.e.,

$$\mathcal{P}_{L-1}(\Omega) \subset S_j^{\mathrm{loc}} := \operatorname{span} \Phi_j \subset L_2^{\mathrm{loc}}(\mathbb{R}^n), \tag{1.12}$$

($\mathcal{P}_r(\Omega)$ denoting the set of algebraic polynomials of degree r at most, restricted to Ω). The estimate (1.11) indicating the approximation properties of $\mathcal{S}$, is often referred to as *direct inequality.* In (1.12), we have to use S_j^{loc} since $\mathcal{P}_r(\Omega)$ is *not* a subspace of $L_2(\Omega)$ for unbounded domains $\Omega \subset \mathbb{R}^n$ or $\Omega = \mathbb{R}^n$.

This requirement is known to hold for a variety of classical hierarchies of trial spaces of spline or finite element type. In particular it can usually be verified for rather flexible geometries in contrast to techniques that work in the Fourier domain. In particular, when the domain of the multiresolution spaces is a domain in Euclidean space and the bases are compactly supported it is well known that polynomial exactness of order d is sufficient to ensure (1.11) for $d_{\mathcal{S}} = d$ by e.g. a Bramble-Hilbert type argument (see, e.g., [18]).

Bernstein Inequality. While the Jackson estimate is linked to the order of approximation of the system, the following inequality refers to the smoothness, [47]. Again, this is a property of the multiresolution spaces. Let us define

$$t := \sup\{t' \in \mathbb{R} : \Phi_j \subset H^{t'}(\Omega)\}. \tag{1.13}$$

Then, the announced inequality reads

$$\|v_j\|_{s,\Omega} \lesssim 2^{js} \|v_j\|_{0,\Omega}, \quad v_j \in S_j, \; s < \gamma_{\mathcal{S}}, \tag{1.14}$$

where of course $\gamma_{\mathcal{S}} \leq t$. This inequality is also known for many concrete examples and is often called *inverse estimate.*

The Characterization Theorem. Jackson and Bernstein inequalities already allow the characterization of certain Sobolev spaces in terms of associated projectors. This is due to the following result, [44, 45]:

Theorem 1. *Let $\mathcal{S}$ be a dense nested sequence of closed subspaces of $L_2(\Omega)$. Let $\mathcal{P} := \{P_j\}$ be an associated sequence of uniformly bounded projectors satisfying $P_j P_l = P_j$ holds for $j \leq l$.*

Moreover, assume that $\mathcal{S}$ and the range $\tilde{\mathcal{S}}$ of $\mathcal{P}^* := \{P_j^*\}$ (the adjoints of P_j) satisfy Jackson- and Bernstein inequalities for $d := d_{\mathcal{S}}$, $\tilde{d} := d_{\tilde{\mathcal{S}}}$ and

$$0 < \gamma := \min\{\gamma_{\mathcal{S}}, d\}, \qquad 0 < \tilde{\gamma} := \min\{\gamma_{\tilde{\mathcal{S}}}, \tilde{d}\},$$

respectively.

Then

$$\sum_{j=0}^{\infty} 2^{2sj} \|(P_j - P_{j-1})f\|_{0,\Omega}^2 \sim \|f\|_{s,\Omega}^2, \qquad s \in (-\tilde{\gamma}, \gamma), \tag{1.15}$$

where it is to be understood that $H^s(\Omega) = (H^{-s}(\Omega))'$ for $s < 0$. Moreover, $\mathcal{P}$ is uniformly bounded in $H^s(\Omega)$ for that range:

$$\|P_j v\|_{s,\Omega} \lesssim \|v\|_{s,\Omega}, \qquad v \in H^s(\Omega), \quad s \in (-\tilde{\gamma}, \gamma). \qquad \square$$

1.1.3 Riesz Basis Property in $L_2(\Omega)$

If one aims to prove the Riesz basis property for a wavelet basis in $L_2(\Omega)$, it turns out that the norm equivalence (1.4) (p. 2) is in fact easier to verify for $H^{|s|}(\Omega)$, provided that the basis functions have the desired smoothness.

Biorthogonality. Let us now assume for a moment that Ψ is a Riesz basis in $L_2(\Omega)$ (i.e., (1.4) holds for $H = L_2(\Omega)$ and $\boldsymbol{D} = \boldsymbol{I}$). Then, the Riesz Representation Theorem implies the existence of a *dual* wavelet basis

$$\tilde{\Psi} := \{\tilde{\psi}_\lambda : \lambda \in \mathcal{J}\} \subset L_2(\Omega), \tag{1.16}$$

i.e.,

$$(\psi_\lambda, \tilde{\psi}_{\lambda'})_{0,\Omega} = \delta_{\lambda,\lambda'}, \qquad \lambda, \lambda' \in \mathcal{J}.$$

By a simple duality argument one can show that also $\tilde{\Psi}$ is a Riesz basis for $L_2(\Omega)$, i.e.,

$$\Big\| \sum_{\lambda \in \mathcal{J}} \tilde{d}_\lambda \tilde{\psi}_\lambda \Big\|_{0,\Omega} \sim \|\tilde{\boldsymbol{d}}\|_{\ell_2(\mathcal{J})}, \qquad \tilde{\boldsymbol{d}} := \{\tilde{d}_\lambda\}_{\lambda \in \mathcal{J}}. \tag{1.17}$$

Thus, biorthogonality is a necessary condition for the Riesz basis property in $L_2(\Omega)$. Let us now collect all facts of dual multiresolution and biorthogonal wavelets. This can then be combined with the characterization statement in Theorem 1 to obtain a wavelet characterization of Sobolev- and Besov spaces.

Biorthogonal Expansions. With the aid of $\tilde{\Psi}$, the expansion coefficients d_λ and $\tilde{d}_\lambda$ of two functions

$$f = \sum_{\lambda \in \mathcal{J}} d_\lambda \, \psi_\lambda, \quad \tilde{f} = \sum_{\lambda \in \mathcal{J}} \tilde{d}_\lambda \, \tilde{\psi}_\lambda \in L_2(\Omega),$$

with respect to the wavelet bases Ψ and $\tilde{\Psi}$, respectively, can be written as

$$d_\lambda = (f, \tilde{\psi}_\lambda)_{0,\Omega}, \qquad \tilde{d}_\lambda = (\tilde{f}, \psi_\lambda)_{0,\Omega}, \qquad \lambda \in \mathcal{J}. \tag{1.18}$$

The pair Ψ, $\tilde{\Psi}$ is often referred to as *biorthogonal system.* Note that the case of orthonormal wavelets is a special case of the above setting.

Dual Multiresolution. We consider $\tilde{\Psi}$ to be generated by a second multiresolution analysis $\tilde{\mathcal{S}} := \{\tilde{S}_j\}_{j \in \mathbb{Z}}$, $\tilde{S}_j := S(\tilde{\Phi}_j)$, $\tilde{\Phi}_j = \{\tilde{\varphi}_{j,k};\ k \in \mathcal{I}_j\}$, which is *dual* to $\mathcal{S}$ in the sense

$$(\varphi_{j,k}, \tilde{\varphi}_{j,k'})_{0,\Omega} = \delta_{k,k'}, \tag{1.19}$$

for $j \in \mathbb{Z}$ and $k, k' \in \mathcal{I}_j$. For the dual system, we denote the refinement matrix by $\tilde{M}_{j,0} \in [\ell_2(\mathcal{I}_j), \ell_2(\mathcal{I}_{j+1})]$, i.e.,

$$\tilde{\Phi}_j = \tilde{\boldsymbol{M}}_{j,0}^T \tilde{\Phi}_{j+1}. \tag{1.20}$$

Dual Wavelets. Again, we can define complement spaces $\tilde{W}_j := \tilde{S}_{j+1} \ominus \tilde{S}_j$. Now, we fix the complements by the condition

$$S_j \perp \tilde{W}_j, \ \tilde{S}_j \perp W_j, \tag{1.21}$$

in the sense $(\Phi_j, \tilde{\Psi}_j)_{0,\Omega} = \langle \tilde{\Phi}_j, \Psi_j \rangle = 0$ (where we use the short-hand notation $(\Theta, \Phi)_{0,\Omega} := \big((\theta, \phi)_{0,\Omega}\big)_{\theta\in\Theta,\phi\in\Phi}$ for two countable systems of functions).

Since $W_j \subset S_{j+1}$, $\tilde{W}_j \subset \tilde{S}_{j+1}$, one has to determine

$$M_{j,1} := (m^j_{k',k})_{k'\in\mathcal{I}_{j+1},k\in\mathcal{J}_j} \in [\ell_2(\mathcal{J}_j), \ell_2(\mathcal{I}_{j+1})],$$
$$\tilde{M}_{j,1} := (\tilde{m}^j_{k',k})_{k'\in\mathcal{I}_{j+1},k\in\mathcal{J}_j} \in [\ell_2(\mathcal{J}_j), \ell_2(\mathcal{I}_{j+1})]$$

and

$$\Psi_j = M^T_{j,1}\Phi_{j+1}, \quad \tilde{\Psi}_j = \tilde{\boldsymbol{M}}^T_{j,1}\tilde{\Phi}_{j+1}, \tag{1.22}$$

such that

$$(\psi_{j,k}, \tilde{\psi}_{j,k'})_{0,\Omega} = \delta_{k,k'}, \qquad j \in \mathbb{Z},\ k, k' \in \mathcal{J}_j. \tag{1.23}$$

Biorthogonal Projectors. Let us now assume for a moment that Ψ, $\tilde{\Psi}$ is a system of biorthogonal wavelets induced by the single scale systems Φ, $\tilde{\Phi}$. These functions induce (biorthogonal) projectors that we have already seen in Theorem 1 on p. 6 above. For any $\Lambda \subset \mathcal{J}$, we define for $f \in H$, $\tilde{f} \in H'$

$$Q_\Lambda f := \sum_{\lambda\in\Lambda} (f, \tilde{\psi}_\lambda)_{0,\Omega}\, \psi_\lambda, \qquad \tilde{Q}_\Lambda \tilde{f} := \sum_{\lambda\in\Lambda} (\tilde{f}, \psi_\lambda)_{0,\Omega}\, \tilde{\psi}_\lambda. \tag{1.24}$$

The projectors corresponding to the index sets $\mathcal{J}_j$ will be denoted by Q_j and $\tilde{Q}_j$, respectively. We will denote the biorthogonal projectors onto S_j by P_j and the following relations hold by (1.21)

$$P_j f = \sum_{|\lambda|<j} (f, \tilde{\psi}_\lambda)_{0,\Omega}\, \psi_\lambda = \sum_{k\in\mathcal{I}_j} (f, \tilde{\varphi}_{j,k})_{0,\Omega}\, \varphi_{j,k}, \quad f \in L_2(\Omega), \tag{1.25}$$

as well as

$$\tilde{P}_j \tilde{f} = \sum_{k\in\mathcal{I}_j} (\tilde{f}, \varphi_{j,k})_{0,\Omega}\, \tilde{\varphi}_{j,k}, \quad \tilde{f} \in L_2(\Omega).$$

It can readily be seen that these operators are in fact projectors and that $P^*_j = \tilde{P}_j$, where F^* is the *adjoint* of an operator F.

1.1.4 Norm Equivalences

Theorem 1 precisely describes the role of the multiresolution sequences $\mathcal{S}$ and $\tilde{\mathcal{S}}$ for the construction of wavelets. The projectors $Q_j := P_j - P_{j-1}$ can be represented by the complement bases, as we shall see by the following result which was shown in [46] (see also [47]):

Theorem 2. *Let the assumptions of Theorem 1 hold and assume that Ψ_j, $\tilde{\Psi}_j$ are stable bases for W_j, $\tilde{W}_j$, respectively. Then, the following norm equivalences hold*

$$\Big\| \sum_{\lambda \in \mathcal{J}} d_\lambda \psi_\lambda \Big\|_{m,\Omega} \sim \Big(\sum_{\lambda \in \mathcal{J}} 2^{2m|\lambda|} |d_\lambda|^2 \Big)^{1/2}, \tag{1.26}$$

$$\Big\| \sum_{\lambda \in \mathcal{J}} d_\lambda \tilde{\psi}_\lambda \Big\|_{m,\Omega} \sim \Big(\sum_{\lambda \in \mathcal{J}} 2^{2m|\lambda|} |d_\lambda|^2 \Big)^{1/2}, \tag{1.27}$$

for $m \in (-\tilde{\gamma}, \gamma)$. □

The above theorem shows that under the above conditions, Ψ, $\tilde{\Psi}$ are wavelet bases in the sense of Definition 1 (p. 2) for $H^m(\Omega)$, $m \in (-\tilde{\gamma}, \gamma)$. Moreover, we see that the biorthogonal projectors are realized in terms of the biorthogonal wavelets in the sense

$$Q_j = P_{j+1} - P_j = \sum_{k \in \mathcal{J}_j} (f, \tilde{\psi}_{j,k})_{0,\Omega}\, \psi_{j,k}$$

as well as

$$\tilde{Q}_j = Q_j^* = P_{j+1}^* - P_j^* = \sum_{k \in \mathcal{J}_j} (f, \psi_{j,k})_{0,\Omega}\, \tilde{\psi}_{j,k}.$$

Besov Spaces. Similar relations are also known to hold for Sobolev spaces in L_p for $p \neq 2$. Moreover, interpolation between such spaces provides norm equivalences for a whole range of *Besov spaces* $B_q^\alpha(L_p)$ [46, 69, 76, 108]. In the present context we will have to make use of the following special case

$$\|\boldsymbol{d}\|_{\ell_\tau(\mathcal{J})} \sim \Big\| \sum_{\lambda \in \mathcal{J}} d_\lambda\, \psi_\lambda \Big\|_{B_\tau^\alpha(L_\tau)} =: \|\boldsymbol{d}^T \Psi\|_{B_\tau^\alpha(L_\tau)}, \tag{1.28}$$

where the smoothness index α and the integrability index τ are related by

$$\frac{1}{\tau} = \frac{\alpha}{n} + \frac{1}{2}. \tag{1.29}$$

1.1.5 General Setting Continued

Coming back to the general setting in the beginning of this chapter. We usually normalize the wavelets in $L_2(\Omega)$, i.e.,

$$\|\psi_\lambda\|_{0,\Omega} \sim 1,$$

so that we obtain from (1.4) on p. 2 the estimate $\|\boldsymbol{d}^T \Psi\|_{0,\Omega} \sim \|\boldsymbol{d}\|_{\ell_2(\mathcal{J})}$.

Let us now consider a Banach subspace H of $L_2(\Omega)$ and denote by H' the dual of H with $L_2(\Omega)$ as pivot space. Then, assuming that Ψ is a wavelet basis for H, we have

$$\|\boldsymbol{D}\boldsymbol{d}\|_{\ell_2(\mathcal{J})} \sim \|\boldsymbol{d}^T\Psi\|_H, \tag{1.30}$$

which by duality implies

$$\|\boldsymbol{D}^{-T}\boldsymbol{d}\|_{\ell_2(\mathcal{J})} \sim \|\boldsymbol{d}^T\tilde{\Psi}\|_{H'}. \tag{1.31}$$

In fact

$$\begin{aligned}
\|\boldsymbol{d}^T\tilde{\Psi}\|_{H'} &= \sup_{u\in H} \frac{(\boldsymbol{d}^T\tilde{\Psi}, u)_{0,\Omega}}{\|u\|_H} &&= \sup_{u=\mathbf{c}^T\Psi\in H} \frac{\boldsymbol{d}^T(\tilde{\Psi},\Psi)_{0,\Omega}\,\mathbf{c}}{\|\mathbf{c}^T\Psi\|_H} \\
&\sim \sup_{\mathbf{c}\in\ell_2(\mathcal{J})} \frac{\boldsymbol{d}^T\mathbf{c}}{\|\boldsymbol{D}\mathbf{c}\|_{\ell_2(\mathcal{J})}} &&= \sup_{\mathbf{c}\in\ell_2(\mathcal{J})} \frac{\boldsymbol{d}^T\boldsymbol{D}^{-1}\mathbf{c}}{\|\mathbf{c}\|_{\ell_2(\mathcal{J})}} \\
&= \sup_{\mathbf{c}\in\ell_2(\mathcal{J})} \frac{(\boldsymbol{D}^{-T}\boldsymbol{d})^T\mathbf{c}}{\|\mathbf{c}\|_{\ell_2(\mathcal{J})}} &&= \|\boldsymbol{D}^{-T}\boldsymbol{d}\|_{\ell_2(\mathcal{J})}.
\end{aligned}$$

Finite Subsets: Adaptive Wavelet Spaces. Using wavelet bases in a Galerkin approach for numerically solving certain operator equations (see also Sect. 3.1.1 on p. 109 below) requires to construct *finite* dimensional test and trial spaces. Here, this will be performed by selecting a finite subset

$$\Lambda \subset \mathcal{J}, \qquad \#\Lambda < \infty,$$

which may result from a refinement strategy or incorporate a-priori knowledge about the solution. The corresponding spaces will be denoted by

$$S_\Lambda := S(\Psi_\Lambda) := \operatorname{span}\Psi_\Lambda, \qquad \tilde{S}_\Lambda := S(\tilde{\Psi}_\Lambda) := \operatorname{span}\tilde{\Psi}_\Lambda, \tag{1.32}$$

where

$$\Psi_\Lambda := \{\psi_\lambda : \lambda \in \Lambda\}, \qquad \tilde{\Psi}_\Lambda := \{\tilde{\psi}_\lambda : \lambda \in \Lambda\}.$$

Sometimes, we will also consider the *full* (i.e., non-adaptive) spaces induced by the index sets

$$\mathcal{J}_j := \{\mu \in \mathcal{J} : |\mu| = j\}, \qquad \mathcal{J}_{[j]} := \{\mu \in \mathcal{J} : |\mu| \le j\}. \tag{1.33}$$

1.1.6 Further Wavelet Features

Besides the characterization of Sobolev- and Besov spaces, wavelets offer at least two more features that will be needed in the sequel.

Locality. Wavelet expansions historically arose from many different fields, see e.g. [63]. The first constructions of wavelets on the real line gave rise to exponentially decaying functions. The advantage of these wavelets is that this kind of locality is also true for the Fourier transform which is used in particular for applications in signal and image processing.

When one aims to use wavelets as trial and test functions for a Galerkin approximation of a partial differential equation, exponential decay is somewhat problematic since truncations have to be performed. This is the reason why we focus on compactly supported wavelets only. Starting with Daubechies' orthonormal wavelet bases [63], the construction of compactly supported wavelet bases on $\mathbb{R}$ has been considered by many authors. The dilation (see also Sect. 1.2 below) implies then that the size of the wavelet support decreases with increasing level in the sense of (1.5) on p. 3. When constructing wavelet bases on more complex domains one has to perform the construction in such a way that (1.5) holds.

Cancellation Property. Another main requirement on the wavelet bases is that integration of a function against a wavelet annihilates the smooth part of the function, in the sense that

$$|(v, \psi_\lambda)_{0,\Omega}| \lesssim 2^{-|\lambda|(\tilde{m}+\frac{n}{2})} \|v\|_{W_\infty^{\tilde{m}}(\operatorname{supp}\psi_\lambda)}, \tag{1.34}$$

where the positive integer $\tilde{m}$ is related to the dual basis $\tilde{\Psi}$ and n is the spatial dimension of the domain ψ_λ is defined on. In the classical case $\tilde{m}$ is the order of *vanishing polynomial moments*, see [47]. Property (1.34) ensures that matrix representations are almost sparse for a wide class of operators, [16, 47]. This is in particular important for integral operators since the corresponding stiffness matrices are densely populated, [16, 36, 47].

1.1.7 A Program for Constructing Wavelets

Let us finally add some comments on the way wavelets are usually constructed. We will follow this path when constructing wavelet bases for different instances of Ω in the sequel. The starting point for a wavelet construction is usually a (primal) multiresolution analysis $\mathcal{S}$. The spaces S_j need to satisfy Jackson and Bernstein inequalities. Then, one has to construct (or take a known) dual multiresolution analysis $\tilde{\mathcal{S}}$ and to ensure Jackson and Bernstein inequalities also for $\tilde{\mathcal{S}}$.

Finally, it remains to construct functions Ψ, $\tilde{\Psi}$ that are Riesz bases for W_j, $\tilde{W}_j$, respectively, such that the constants in the Riesz basis estimates are independent of the level j. Constructing these complement bases can sometimes be done 'by hands' by computing $M_{j,1}$, $\tilde{M}_{j,1}$ in (1.22) on p. 8 directly. However, sometimes this may become a delicate task. A general program is given in [25] called *stable completion*. The idea is to construct an initial stable completion of $\mathcal{S}$ in the sense that one constructs a basis $\check{\Psi}_j$ of *any* complement space $\check{W}_j$ of S_j in S_{j+1} such that

$$\|\boldsymbol{c}_j^T \check{\Psi}_j\|_H \sim \|\boldsymbol{c}_j\|_{\ell_2(\mathcal{J}_j)}$$

with constants independent of j. This is sometimes relatively easy to perform. In particular, it allows in this step to work with the primal system S_j only. Then, these bases are projected appropriately onto the desired complement spaces W_j determined by $\tilde{\mathcal{S}}$. We will use this procedure in Sect. 1.3 on p. 17 below.

Note that it is not always necessary to have the dual multiresolution analysis $\tilde{\mathcal{S}}$ at hand *explicitly*. Biorthogonality can also be ensured by determining the projectors P_j in (1.25) on p. 8 in such a way that

$$P_j\, P_{j+1} = P_j.$$

Stability with constants independent of j can be shown by proving $\|P_j\| \lesssim 1$ for these projectors, [47].

One may also view the construction of wavelet systems as follows, [47]: Given a rectangular matrix $\boldsymbol{M}_{j,0} = (m^j_{k',k})_{k'\in\mathcal{I}_{j+1},k\in\mathcal{I}_j}$, find a sparse completion $\boldsymbol{M}_{j,1} = (m^j_{k',k})_{k'\in\mathcal{I}_{j+1},k\in\mathcal{J}_j}$ such that the composed matrix $\boldsymbol{M}_j := (\boldsymbol{M}_{j,0}\ \boldsymbol{M}_{j,1})$ is invertible, its inverse is also sparse and one has

$$\|\boldsymbol{M}_j\|, \|\boldsymbol{M}_j^{-1}\| = \mathcal{O}(1), \quad j \to \infty.$$

1.2 Wavelets on the Real Line

The easiest situation occurs when one considers wavelets on the real line. Here, scaling and wavelet systems are usually constructed by dyadic dilations and translations of appropriate generators and mother wavelets. Since later we will be concerned with tensor products of univariate scaling and wavelet systems, it will be convenient to use a different notation for the systems in 1D, i.e, we denote scaling functions and systems by ξ and Ξ, respectively, and wavelets and wavelet systems by η and Υ, respectively.

A dual pair of generators ξ, $\tilde{\xi} \in L_2(\mathbb{R})$ is given by their refinement equations

$$\xi(x) = \sum_{k\in\mathbb{Z}} a_k\, \xi(2x-k), \qquad \tilde{\xi}(x) = \sum_{k\in\mathbb{Z}} \tilde{a}_k\, \tilde{\xi}(2x-k) \tag{1.35}$$

and the duality relation

$$(\xi, \tilde{\xi}(\cdot - k))_{0,\mathbb{R}} = \delta_{0,k}, \qquad k \in \mathbb{Z}. \tag{1.36}$$

As already mentioned, such functions are also called *scaling functions*. The coefficients $\{a_k\}_{k\in\mathbb{Z}}$, $\{\tilde{a}_k\}_{k\in\mathbb{Z}}$ are called *masks* of ξ and $\tilde{\xi}$, respectively. Since we are ultimately interested in local functions, we only consider finite masks, i.e.,

$$a_k \neq 0 \iff N_1 \leq k \leq N_2; \qquad \tilde{a}_k \neq 0 \iff \tilde{N}_1 \leq k \leq \tilde{N}_2$$

for some $N_1, N_2, \tilde{N}_1, \tilde{N}_2 \in \mathbb{N}$.

A pair of (biorthogonal) wavelets η, $\tilde{\eta}$ is then defined by the relations

$$\eta(x) = \sum_{k\in\mathbb{Z}} b_k\, \xi(2x-k), \qquad \tilde{\eta}(x) = \sum_{k\in\mathbb{Z}} \tilde{b}_k\, \tilde{\xi}(2x-k)$$

as well as

$$(\eta, \tilde{\eta}(\cdot - k))_{0,\mathbb{R}} = \delta_{0,k}, \quad (\xi, \tilde{\eta}(\cdot - k))_{0,\mathbb{R}} = (\tilde{\xi}, \eta(\cdot - k))_{0,\mathbb{R}} = 0, \qquad k \in \mathbb{Z}.$$

The wavelet coefficients b_k, $\tilde{b}_k$ are determined by the mask coefficients in the following way, [30, 63]

$$b_k = (-1)^k\, \tilde{a}_{1-k}, \quad \tilde{b}_k = (-1)^k\, a_{1-k}. \qquad k \in \mathbb{Z}.$$

Then, for any $j \in \mathbb{Z}$, the primal scaling systems $\Xi_j^{\mathbb{R}} = \{\xi_{[j,k]} : k \in \mathbb{Z}\}$ are defined by the relations

$$\xi_{[j,k]}(x) := 2^{j/2}\xi(2^j x - k); \tag{1.37}$$

similar definitions hold for the dual scaling systems $\tilde{\Xi}_j^{\mathbb{R}}$ and for the wavelet systems $\Upsilon_j^{\mathbb{R}}$, $\tilde{\Upsilon}_j^{\mathbb{R}}$.

Let us now describe some examples. The list of presented examples is far from being complete. We concentrate on those systems that we will refer to in the sequel.

1.2.1 Orthonormal Wavelets

Scaling Functions. In this case, one has $\xi = \tilde{\xi}$. For each $N \in \mathbb{N}$, there exists one scaling function ${}_N\xi$ supported in $[0, 2N-1]$ such that the generated wavelet ${}_N\eta$ (defined by the wavelet coefficients $b_k := (-1)^k a_{1-k}$) has a maximal number of vanishing moments, namely $2N$, [63], pp. 194. The orthonormal scaling function for $N = 1$ is the characteristic function $\chi_{[0,1)}$. The orthonormal scaling functions for $N = 2, 3, 4, 5$ are displayed in Fig. 1.1.

Wavelets. For each orthonormal scaling function ${}_N\xi$, the corresponding wavelet ${}_N\eta$ is determined by their wavelet coefficients by $b_k = (-1)^k a_{1-k}$. The orthonormal wavelet corresponding to ${}_1\xi = \chi_{[0,1)}$ is the *Haar-Wavelet* ${}_1\eta = \frac{1}{2}\chi_{[0,0.5)} - \frac{1}{2}\chi_{[0.5,1]}$. The orthonormal wavelets ${}_N\eta$ for $N = 2, 3, 4, 5$ are displayed in Fig. 1.2 on p. 14.

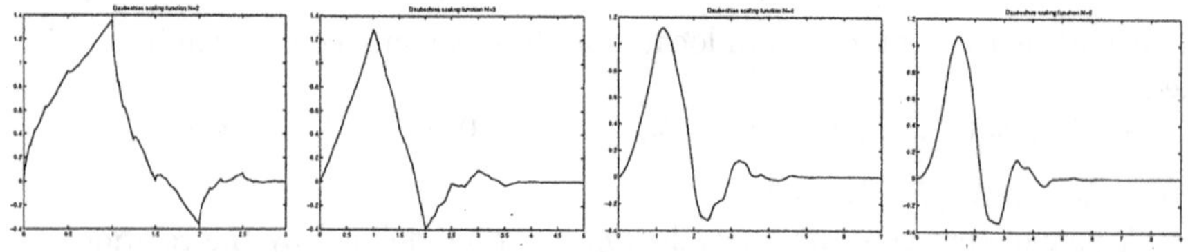

Fig. 1.1. Daubechies' orthogonal scaling function ${}_N\xi$ for $N = 2, 3, 4, 5$.

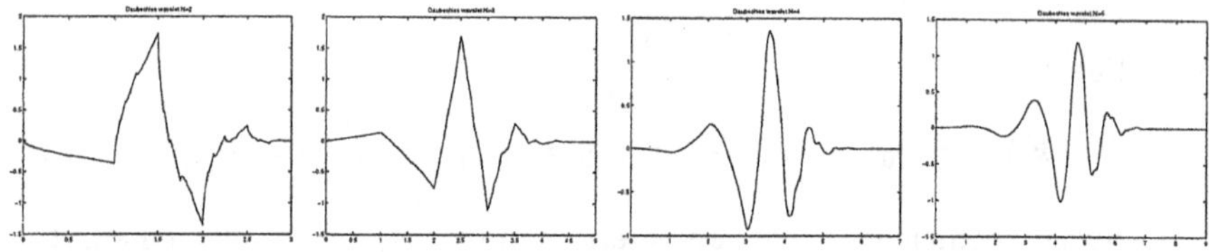

Fig. 1.2. Daubechies' orthogonal wavelets ${}_N\eta$ for $N = 2, 3, 4, 5$.

1.2.2 Biorthogonal B-Spline Wavelets

Primal Scaling Functions. Let us denote for a sequence of knots $t_i \leq \ldots \leq t_{i+d}$ by $[t_i, \ldots, t_{i+d}]f$ the d–th order divided difference of $f \in C^d(\mathbb{R})$ at $t_i, \ldots, t_{i+d}$. Setting $x_+^d := (\max\{0, x\})^d$, the *cardinal B–spline* ${}_d\xi$ of order $d \in \mathbb{N}$ is defined as

$$
{}_d\xi(x) := d\,[0, 1, \ldots, d]\,\left(x + \lfloor \tfrac{d}{2} \rfloor\right)_+^{d-1}. \tag{1.38}
$$

These functions are displayed in the first column of Fig. 1.3 on p. 16 for $d = 1, 2, 3$.

Dual Scaling Functions. It has been shown in [30] that for each d and any $\tilde{d} \geq d$, $\tilde{d} \in \mathbb{N}$, so that $d + \tilde{d}$ is even, there exists a compactly supported refinable function ${}_{d,\tilde{d}}\tilde{\xi} \in L_2(\mathbb{R})$ whose regularity (and support length) increases proportionally with $\tilde{d}$ and which is dual to ${}_d\xi$. The functions corresponding to the choices $d = 1$, $\tilde{d} = 5$ as well as $d = 2$, $\tilde{d} = 4$ and $d = \tilde{d} = 3$ are shown in the second column of Fig. 1.3. The refinement coefficients can be found in [30].

Since we will use these functions frequently in the sequel, let us summarize their most important properties. First of all, cardinal B-splines are symmetric, i.e.,

$$
{}_d\xi(x + \mu(d))) = {}_d\xi(-x), \qquad x \in \mathbb{R}, \tag{1.39}
$$

where $\mu(d) := d \bmod 2$, and have support

$$
\operatorname{supp} {}_d\xi = \left[\tfrac{1}{2}(-d + \mu(d)), \tfrac{1}{2}(d + \mu(d))\right] = \left[-\lfloor \tfrac{d}{2} \rfloor, \lceil \tfrac{d}{2} \rceil\right] =: [\ell_1, \ell_2], \tag{1.40}
$$

i.e., $d = \ell_2 - \ell_1$ and $\mu(d) = \ell_1 + \ell_2$. Thus, the B-splines of even order are centered around 0 while the ones of odd order are symmetric around $\frac{1}{2}$. Moreover, they are refinable

$$ {}_d\xi(x) = \sum_{k=\ell_1}^{\ell_2} 2^{1-d} \binom{d}{k + \lfloor \frac{d}{2} \rfloor} {}_d\xi(2x-k) =: \sum_{k=\ell_1}^{\ell_2} a_k \, {}_d\xi(2x-k). \tag{1.41} $$

Finally, ${}_d\xi$ is exact of order d, i.e., $\mathcal{P}_{d-1} \subset \mathrm{span}({}_d\Xi_j^{\mathbb{R}})$ for all j. The dual scaling functions have the following properties:

(i) ${}_{d,\tilde{d}}\tilde{\xi}$ has compact support,

$$ \mathrm{supp}\, {}_{d,\tilde{d}}\tilde{\xi} = [\ell_1 - \tilde{d} + 1, \ell_2 + \tilde{d} - 1] =: [\tilde{\ell}_1, \tilde{\ell}_2]. \tag{1.42} $$

(ii) ${}_{d,\tilde{d}}\tilde{\xi}$ is refinable with finitely supported mask $\tilde{\mathbf{a}} = \{\tilde{a}_k\}_{k=\tilde{\ell}_1}^{\tilde{\ell}_2}$, see [30].

(iii) ${}_{d,\tilde{d}}\tilde{\xi}$ is symmetric i.e., ${}_{d,\tilde{d}}\tilde{\xi}(x + \mu(d)) = {}_{d,\tilde{d}}\tilde{\xi}(-x)$, $x \in \mathbb{R}$.

(iv) ${}_{d,\tilde{d}}\tilde{\xi}$ is exact of order $\tilde{d}$, i.e., all polynomials of degree less than $\tilde{d}$ can be represented as linear combinations of the translates ${}_{d,\tilde{d}}\tilde{\xi}(\cdot - k), k \in \mathbb{Z}$.

(v) The regularity of ${}_{d,\tilde{d}}\tilde{\xi}$ increases proportionally with $\tilde{d}$.

One easily checks that the symmetry properties (1.39) and (iii), have the following discrete counterparts

$$ a_k = a_{\mu(d)-k}, \quad \tilde{a}_k = \tilde{a}_{\mu(d)-k}, \quad k \in \mathbb{Z}. \tag{1.43} $$

Wavelets. For each pair ${}_d\xi$, ${}_{d,\tilde{d}}\tilde{\xi}$ of dual B–spline generators, the corresponding biorthogonal wavelets ${}_{d,\tilde{d}}\eta$, ${}_{d,\tilde{d}}\tilde{\eta}$ can be found in [30]. Their mask coefficients are given by $b_k = (-1)^k \tilde{a}_{1-k}$ and $\tilde{b}_k = (-1)^k a_{1-k}$. The primal wavelets are displayed in the third column of Fig. 1.3 for $d = 1$, $\tilde{d} = 5$, $d = 2$, $\tilde{d} = 4$ and $d = \tilde{d} = 3$. The corresponding dual wavelets are contained in the fourth column.

1.2.3 Interpolatory Wavelets

Interpolatory wavelets will be needed in the sequel since the expansion coefficients are obtained by sampling rather than by integration. One prominent example are *hierarchical bases*, [119]. We follow here mostly the description of [70].

Definition 2. *A compactly supported function ξ is called* (r, d)-interpolatory scaling function *if*

a.) it interpolates at the integer nodes, i.e., $\xi(k) = \delta_{0,k}$, $k \in \mathbb{Z}$;

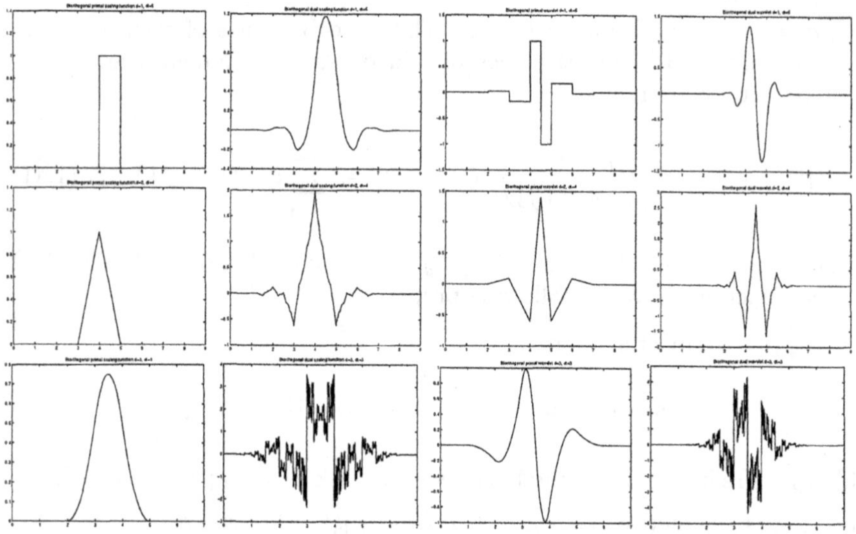

Fig. 1.3. Biorthogonal B-spline scaling functions and wavelets for $d = 1$, $\tilde{d} = 5$ (first row), $d = 2$, $\tilde{d} = 4$ (second row) and $d = \tilde{d} = 3$ (third row). The first column shows the primal scaling function ${}_d\xi$, the second the dual scaling function ${}_{d,\tilde{d}}\tilde{\xi}$ whereas the third and fourth column contain the primal and dual wavelets ${}_{d,\tilde{d}}\eta$ and ${}_{d,\tilde{d}}\tilde{\eta}$, respectively.

b.) it fulfills a two-scale relation

$$\xi(x) = \sum_k \xi(k/2)\,\xi(2x - k);$$

c.) $\mathcal{P}_d(\mathbb{R}) \subset S_j^{\mathrm{loc}}$*;*
d.) ξ *is* Hölder continuous *of order* r. □

There are two well-known families of such functions. The first one are interpolatory splines of degree d yielding an $(d-1, d)$-interpolatory wavelet. The simplest example of this family is $\xi = {}_2\xi$, i.e., the piecewise linear B-spline. The corresponding scaling basis (here on $(0,1)$) is displayed in the left picture in Fig. 1.4. The second family consists of the *Deslaurier-Debuc* fundamental functions, [65]. The induced spaces are again nested.

Then, setting

$$\eta(x) := \xi(2(x - \tfrac{1}{2}))$$

gives rise to an (r, d)-*interpolatory wavelet.* The functions η induced by ${}_2\xi$ are shown on the right in Fig. 1.4. Then, for each $f \in C^0(\mathbb{R}) \cap \mathcal{P}_d(\mathbb{R})$, one obtains a wavelet transform, i.e., coefficients $\beta_{j_0,k}$ (j_0 being the coarsest level) and $\alpha_{j,k}$ depending only on the samples of f at level $j+1$ and coarser, such that

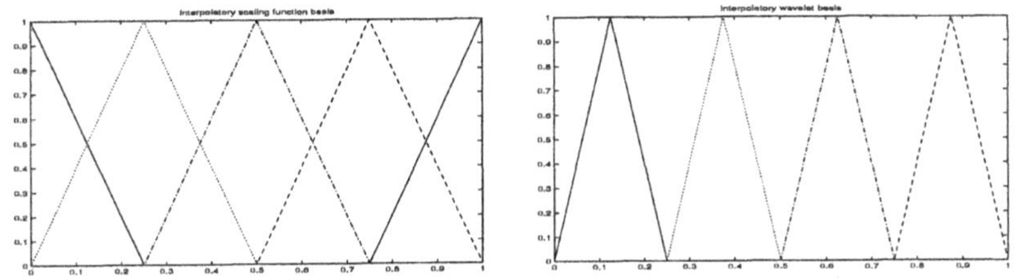

Fig. 1.4. Piecewise linear interpolatory scaling system (left) and wavelet system (right).

$$f = \sum_{k\in\mathbb{Z}} \beta_{j_0,k}\, \xi_{j_0,k} + \sum_{j\geq j_0} \sum_{k\in\mathbb{Z}} \alpha_{j,k}\, \eta_{j,k},$$

where the above sum is convergent in the supremum norm, [70].

Let us mention that also interpolatory wavelets allow to characterize certain Sobolev- and Besov-spaces. It was proven in [70] that for $\min(r,d) > \sigma > 1/p$, $p,q \in (0,\infty]$, one has

$$\|f\|_{B^\sigma_q(L_p(\mathbb{R}))} \sim \|(\beta_{j_0,\cdot})\|_{\ell_p} + \left(\sum_{j=j_0}^{\infty} \left(2^{js}\left(\sum_{k\in\mathbb{Z}} |\alpha_{j,k}|^p\right)^{1/p}\right)^q\right)^{1/q}$$

for $s = \sigma + \frac{1}{2} - \frac{1}{p}$, i.e., they are wavelet bases for these spaces in the sense of Definition 1 on p. 2. Hence, for $p = 2$, we obtain that properly scaled (r,d)-interpolatory wavelets form a Riesz basis for $B^\sigma_q(L_2(\mathbb{R}))$ for $\min(r,d) > \sigma > 1/2$.

Note that $\sigma > 0$, so that negative ranges are *not* covered. Moreover, it was shown in [70] that the bound $\sigma > 1/p$ is in fact sharp. This means that interpolatory wavelets do not form a Riesz basis for $L_2(\Omega)$, which can also be seen from the fact that the sampling operator is not well-defined on $L_2(\Omega)$.

1.3 Wavelets on the Interval

In this section, we describe the construction of wavelet bases on the unit interval $(0,1)$, where we follow the lines in [54]. The main idea behind most constructions in the literature, see e.g. [31, 58, 82], can be summarized as follows: Most of the scaling functions and wavelets on $\mathbb{R}$ whose support is contained in $(0,1)$ are retained, whereas near the boundaries certain linear combinations of restrictions of basis functions on the real line are formed. The ultimate goal is to preserve the important properties of the wavelets namely biorthogonality, locality, vanishing moments and approximation properties ensuring in particular norm equivalences.

Before going into the details, let us describe two naive approaches which in general do *not* lead to the desired result. They describe the intrinsic problems that arise when one aims at generalizing the shift-invariant setting on $\mathbb{R}^n$ to bounded domains.

- One could construct spaces by restricting all those functions to $(0, 1)$ whose support intersect the interval.
- Since in many cases of interest one also has to incorporate boundary conditions, one could consider only those functions whose support is contained in $(0, 1)$ in order to enforce homogeneous Dirichlet boundary conditions.

Both situations are illustrated in Fig. 1.5. As already mentioned, both

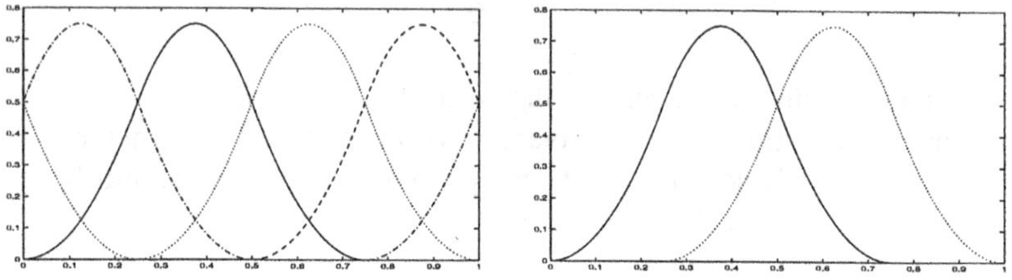

Fig. 1.5. Scaling systems on the interval obtained by restrictions (left) and interior functions (right).

approaches fail. As for the first approach, duality on the real line does not imply the duality of the restrictions. Moreover, since at least for B-splines, the support of the dual functions is larger then those of the primals, a stability problem for the dual functions induced by small overlapping pieces occurs. For the second approach there is a lack of approximation power near the boundaries. Hence a more sophisticated approach is needed in order to preserve duality, stability and approximation properties.

We are now going to sketch the main ingredients of the construction in [54]. The road map is as follows: In Sect. 1.3.1 on p. 19, we define boundary scaling functions in order to preserve the approximation power given by the corresponding functions on the real line. We derive the two-scale relation in a general setting. In Sect. 1.3.2 on p. 20 we consider biorthogonal systems induced by cardinal B-splines based upon the dual B-spline systems in [30]. In order to obtain biorthogonality, in particular the dimensions of Ξ_j and $\tilde{\Xi}_j$ have to match which is enforced in Sect. 1.3.2 on p. 20. Having two scaling systems on the interval at hand, the biorthogonality is usually destroyed by introducing modified functions near the boundaries. Hence one has to show that these modified systems are biorthogonalizable which for B-splines scaling functions is done in Sect. 1.3.3 on p. 25. The resulting refinement matrices are presented in Sect. 1.3.4 on p. 30. Sect. 1.3.5 on p. 33 contains the construction of wavelets featuring a general construction principle called *stable completion*, see [25] and also Sect. 1.1.7 on p. 11 above. Finally, in Sect. 1.3.7 on p. 44, we review the realization of different boundary conditions and in Sect. 1.3.8 on p. 45 we add some comments on the extension of the presented construction to other biorthogonal systems than B-splines.

Let us also point out that filter coefficients and software tools for the subsequent construction are available, e.g. [6].

1.3.1 Boundary Scaling Functions

As we already pointed out, we need to build appropriate linear combinations of the original translates in order to adapt the bases to $(0,1)$. Let us first consider such combinations for scaling functions. For all realizations the masks in the refinement relations are needed. Hence, we detail the refinement relations for the boundary scaling functions here.

In this section we consider an *arbitrary* scaling functions ξ which is exact of order d, i.e.,

$$x^r = \sum_{k\in\mathbb{Z}} \alpha_{\tilde{\xi},r}(k)\,\xi(x-k), \quad x \in \mathbb{R} a.e., \quad r = 0,\ldots,d-1, \tag{1.44}$$

where $\alpha_{\tilde{\xi},r}(k) := ((\cdot)^r, \tilde{\xi}(\cdot - k))_{0,\mathbb{R}}$. Moreover, the support of ξ will be denoted by $\operatorname{supp}\xi := [\ell_1, \ell_2]$. Then, the refinement relation takes the following form.

Lemma 1. *Suppose that $\ell \geq -\ell_1$ and define boundary scaling functions by*

$$\xi^L_{j,\ell-d+r} := \sum_{m=-\ell_2+1}^{\ell-1} \alpha_{\tilde{\xi},r}(m)\,\xi_{[j,m]}\big|_{\mathbb{R}_+}, \quad r = 0,\ldots,d-1. \tag{1.45}$$

Then, the following refinement relation holds

$$\xi^L_{j,\ell-d+r} = 2^{-(r+1/2)}\left(\xi^L_{j+1,\ell-d+r} + \sum_{m=\ell}^{2\ell+\ell_1-1} \alpha_{\tilde{\xi},r}(m)\,\xi_{[j+1,m]}\right) + \sum_{m=2\ell+\ell_1}^{2\ell+\ell_2-2} \beta_{\tilde{\xi},r}(m)\,\xi_{[j+1,m]}, \qquad r = 0,\ldots,d-1, \tag{1.46}$$

where

$$\beta_{\tilde{\xi},r}(m) := 2^{-1/2}\left(\sum_{q=\lceil\frac{m-\ell_2}{2}\rceil}^{\ell-1} \alpha_{\tilde{\xi},r}(q)\,a_{m-2q}\right). \quad \square \tag{1.47}$$

The proof is somewhat technical but straightforward and can be found in [54].

1.3.2 Biorthogonal Scaling Functions

From now one, we concentrate on B-spline scaling functions and their duals from [30]. In the following $d, \tilde{d}$ will be arbitrary as above but fixed so that we can suppress them as indices and write briefly $\xi, \tilde{\xi}$. Some of the subsequent results are in fact only valid for B-spline induced systems. We will clearly indicate those facts.

Cardinality and Index Sets. The following construction of biorthogonal multiresolution analyses on $(0, 1)$ follows first again familiar lines in that we retain translates of dilated scaling functions $\xi, \tilde{\xi}$ whose supports are fully contained in $(0, 1)$. Since by (1.42) the support of $\tilde{\xi}$ is at least as large than that of ξ, i.e., $\tilde{\ell}_2 \geq \ell_2, \ -\tilde{\ell}_1 \geq -\ell_1$ (even if $\tilde{d} < d$), we consider first the dual collections and fix some integer $\tilde{\ell}$ satisfying

$$\tilde{\ell} \geq \tilde{\ell}_2, \tag{1.48}$$

so that the indices

$$\tilde{I}_j^0 := \{\tilde{\ell}, \ldots, 2^j - \tilde{\ell} - \mu(d)\} \tag{1.49}$$

correspond to translates $\tilde{\xi}_{[j,m]}$ whose support is contained in $(0, 1)$. To preserve polynomial exactness of degree $\tilde{d}-1$ we need $\tilde{d}$ additional basis functions near the left and right end of the interval which will be constructed according to the recipe from Sect. 1.3.1. The corresponding index sets are then

$$\tilde{I}_j^L := \{\tilde{\ell}-\tilde{d}, \ldots, \tilde{\ell}-1\}, \quad \tilde{I}_j^R := \{2^j-\tilde{\ell}+1-\mu(d), \ldots, 2^j-\tilde{\ell}+\tilde{d}-\mu(d)\}. \tag{1.50}$$

We have included here a shift by $-\mu(d)$ in $\tilde{I}_j^R$ to make best possible use of symmetry later, see also Fig. 1.6.

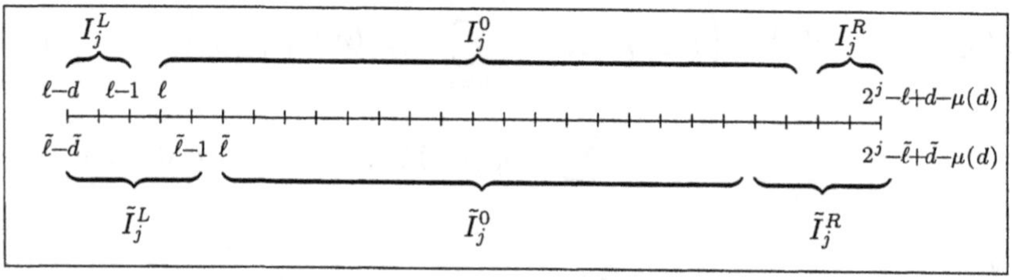

Fig. 1.6. Index sets for the interval for $d = 3$, $\tilde{d} = 5$, $\ell = 4$, $\tilde{\ell} = \tilde{\ell}_2 = 6$.

On the primal side we need bases of the *same cardinality* which causes some adjustments. Since the degree $d - 1$ of exactness is in general less than $\tilde{d} - 1$, the boundary index sets necessarily take the form

$$\begin{aligned} I_j^L &:= \{\tilde{\ell} - \tilde{d}, \ldots, \tilde{\ell} - (\tilde{d} - d) - 1\}, \\ I_j^R &:= \{2^j - \tilde{\ell} + (\tilde{d} - d) + 1 - \mu(d), \ldots, 2^j - \tilde{\ell} + \tilde{d} - \mu(d)\}, \end{aligned} \tag{1.51}$$

so that the *interior translates* $\xi_{[j,m]}$ are determined by $m \in I_j^0$, where

$$I_j^0 := \{\tilde{\ell} - (\tilde{d} - d), \ldots, 2^j - \tilde{\ell} + (\tilde{d} - d) - \mu(d)\}. \tag{1.52}$$

For simplicity, it will always be assumed that j is large enough to ensure that $I_j^0 \neq \emptyset$, i.e., $\tilde{\ell} \leq 2^j - \tilde{\ell} - \mu(d) + 2(\tilde{d} - d)$. For (orthonormal) wavelet bases on coarser scales, we refer to [14]. By construction we have now

$$\tilde{I}_j := \tilde{I}_j^L \cup \tilde{I}_j^0 \cup \tilde{I}_j^R = I_j = I_j^L \cup I_j^0 \cup I_j^R, \tag{1.53}$$

see Fig. 1.6. Moreover, abbreviating

$$\ell := \tilde{\ell} - (\tilde{d} - d), \tag{1.54}$$

we observe that

$$\begin{aligned} I_j^L &= \{\ell - d, \ldots, \ell - 1\}, \\ I_j^R &= \{2^j - \ell + 1 - \mu(d), \ldots, 2^j - \ell + d - \mu(d)\}, \\ I_j^0 &= \{\ell, \ldots, 2^j - \ell - \mu(d)\}, \end{aligned} \tag{1.55}$$

so that primal and dual index sets formally take the same form. Note also that by (1.42) and (1.48), $\ell \geq \tilde{\ell}_2 - (\tilde{d} - d) = -\ell_1 + 2\ell_2 - 1$ so that the functions $\xi_{[j,m]}, m \in I_j^0$, are, under the above assumption on j, indeed supported in $(0, 1)$, i.e.,

$$\operatorname{supp} \xi_{[j,k]} \subset [0,1], \quad k \in I_j^0, \qquad \operatorname{supp} \tilde{\xi}_{[j,k]} \subset [0,1], \quad k \in \tilde{I}_j^0. \tag{1.56}$$

Boundary Scaling Functions. The next step is to construct modified basis functions near the end points of the interval for the primal and dual side following the recipe of Sec. 1.3.1 above. To this end, it will be convenient to abbreviate (see (1.44))

$$\tilde{\alpha}_{m,r} := \alpha_{\tilde{\xi},r}(m), \quad \alpha_{m,r} := \alpha_{\xi,r}(m). \tag{1.57}$$

Since obviously

$$\begin{aligned} \alpha_{j,m,r}^L &:= 2^j \int_{\mathbb{R}} (2^j x)^r \xi(2^j x - m)\, dx = \int_{\mathbb{R}} x^r\, \xi(x - m)\, dx \\ \alpha_{j,m,r}^R &:= 2^j \int_{\mathbb{R}} \left(2^j(1 - x)\right)^r \xi(2^j x - m)\, dx = \int_{\mathbb{R}} (2^j - x)^r \xi(x - m)\, dx, \end{aligned} \tag{1.58}$$

and likewise

$$\tilde{\alpha}^L_{j,m,r} := 2^j \int_{\mathbb{R}} (2^j x)^r \tilde{\xi}(2^j x - m)\, dx = \int_{\mathbb{R}} x^r\, \tilde{\xi}(x-m)\, dx$$
$$\tilde{\alpha}^R_{j,m,r} := 2^j \int_{\mathbb{R}} \left(2^j(1-x)\right)^r \tilde{\xi}(2^j x - m)\, dx = \int_{\mathbb{R}} (2^j - x)^r \tilde{\xi}(x-m)\, dx, \tag{1.59}$$

we conclude on one hand that

$$\sum_{m\in\mathbb{Z}} \alpha^L_{j,m,r} \tilde{\xi}_{[j,m]}(x) = 2^{j(r+1/2)} x^r, \quad \sum_{m\in Z} \alpha^R_{j,m,r} \tilde{\xi}_{[j,m]}(x) = 2^{j(r+1/2)} (1-x)^r, \tag{1.60}$$

for $r = 0, \dots, \tilde{d}-1$ and

$$\sum_{m\in\mathbb{Z}} \tilde{\alpha}^L_{j,m,r} \xi_{[j,m]}(x) = 2^{j(r+1/2)} x^r, \quad \sum_{m\in Z} \tilde{\alpha}^R_{j,m,r} \xi_{[j,m]}(x) = 2^{j(r+1/2)} (1-x)^r,$$

for $r = 0, \dots, d-1$. On the other hand, noting that by the symmetry relation in (1.39) on p. 14,

$$\begin{aligned}\int_{\mathbb{R}} x^r\, \xi(x-m)\, dx &= \int_{\mathbb{R}} (2^j - x)^r\, \xi(2^j - x - m)\, dx \\ &= \int_{\mathbb{R}} (2^j - x)^r\, \xi(x - (2^j - m) + \mu(d))\, dx,\end{aligned}$$

(1.58) also reveals that

$$\begin{aligned}\alpha^L_{j,m,r} &= \alpha_{m,r}, \quad \alpha^R_{j,m,r} = \alpha_{2^j - m - \mu(d), r}, \quad r = 0, \dots, \tilde{d}-1, \\ \tilde{\alpha}^L_{j,m,r} &= \tilde{\alpha}_{m,r}, \quad \tilde{\alpha}^R_{j,m,r} = \tilde{\alpha}_{2^j - m - \mu(d), r}, \quad r = 0, \dots, d-1.\end{aligned} \tag{1.61}$$

Furthermore, recalling the definition of the refinement coefficients in (1.47), let

$$\tilde{\beta}^L_{j,m,r} := \beta_{\tilde{\xi},r}(m), \quad \beta^L_{j,m,r} := \beta_{\xi,r}(m) := 2^{-1/2} \sum_{q=\lceil \frac{m-\tilde{\ell}_2}{2} \rceil}^{\tilde{\ell}-1} \alpha_{q,r}\, \tilde{a}_{m-2q}. \tag{1.62}$$

Employing the symmetry relations for the mask coefficients (1.43) and (1.61), one verifies that

$$\begin{aligned}\beta^L_{j,m,r} &= 2^{-1/2} \sum_{q=2^j-\tilde{\ell}-\mu(d)+1}^{2^j - \lceil \frac{m-\tilde{\ell}_2}{2} \rceil - \mu(d)} \alpha^R_{j,q,r}\, \tilde{a}_{2^{j+1}-m-\mu(d)-2q}, \\ \tilde{\beta}^L_{j,m,r} &= 2^{-1/2} \sum_{q=2^j-\ell-\mu(d)+1}^{2^j - \lceil \frac{m-\ell_2}{2} \rceil - \mu(d)} \tilde{\alpha}^R_{j,q,r}\, a_{2^{j+1}-m-\mu(d)-2q}.\end{aligned} \tag{1.63}$$

Thus defining

$$\begin{aligned}\beta^R_{j,m,r} &:= 2^{-1/2} \sum_{q=2^j-\ell-\mu(d)+1}^{\infty} \alpha^R_{j,q,r}\tilde{a}_{m-2q},\\ \tilde{\beta}^R_{j,m,r} &:= 2^{-1/2} \sum_{q=2^j-\tilde{\ell}-\mu(d)+1}^{\infty} \tilde{\alpha}^R_{j,q,r}a_{m-2q},\end{aligned} \tag{1.64}$$

we easily obtain

$$\beta^R_{j,m,r} = \beta^L_{j,2^{j+1}-m-\mu(d),r}, \quad \tilde{\beta}^R_{j,m,r} = \tilde{\beta}^L_{j,2^{j+1}-m-\mu(d),r}. \tag{1.65}$$

We can now follow the lines of Sect. 1.3.1 to define boundary scaling functions according to (1.45)

$$\begin{aligned}\xi^L_{j,\ell-d+r} &:= \sum_{m=-\ell_2+1}^{\ell-1} \tilde{\alpha}_{m,r}\, \xi_{[j,m]}\big|_{[0,1]},\\ \xi^R_{j,2^j-\ell+d-\mu(d)-r} &:= \sum_{m=2^j-\ell-\mu(d)+1}^{2^j-\ell_1-1} \tilde{\alpha}^R_{j,m,r}\, \xi_{[j,m]}\big|_{[0,1]},\end{aligned} \tag{1.66}$$

for $r = 0, \ldots, d-1$, and likewise on the dual side, i.e.,

$$\begin{aligned}\tilde{\xi}^L_{j,\tilde{\ell}-\tilde{d}+r} &:= \sum_{m=-\tilde{\ell}_2+1}^{\tilde{\ell}-1} \alpha_{m,r}\, \tilde{\xi}_{[j,m]}\Big|_{[0,1]},\\ \tilde{\xi}^R_{j,2^j-\tilde{\ell}+\tilde{d}-\mu(d)-r} &:= \sum_{m=2^j-\tilde{\ell}-\mu(d)+1}^{2^j-\tilde{\ell}_1-1} \alpha^R_{j,m,r}\, \tilde{\xi}_{[j,m]}\Big|_{[0,1]},\end{aligned} \tag{1.67}$$

for $r = 0, \ldots, \tilde{d}-1$, while the interior functions are given by

$$\xi_{j,k} := \xi_{[j,k]}, \quad k \in I^0_j, \qquad \tilde{\xi}_{j,k} := \tilde{\xi}_{[j,k]}, \quad k \in \tilde{I}^0_j. \tag{1.68}$$

The functions $\xi^L_{5,k}$ and $\tilde{\xi}^L_{5,k}$ are displayed for $d = 3$ and $\tilde{d} = 5$ in Figs. 1.7 and 1.8.

Biorthogonal Multiresolution. In the sequel we will always assume that

$$j \geq \left\lceil \log_2(\tilde{\ell} + \tilde{\ell}_2 - 1) + 1 \right\rceil =: j_0 \tag{1.69}$$

so that the supports of the left and right end functions do not overlap (but see Remark 3 on p. 40). However, we point out that this is a mere technical assumption. We refer to [14] for a construction of wavelets on coarse scales. The following symmetry relations will be used frequently.

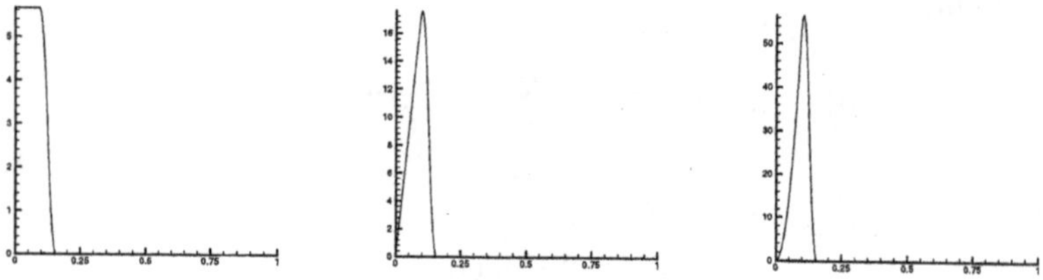

Fig. 1.7. Primal modified single scale functions $\xi^L_{5,k}$ near the left boundary for $d = 3$ and $\tilde{d} = 5$, $k = 1, \ldots, 3$.

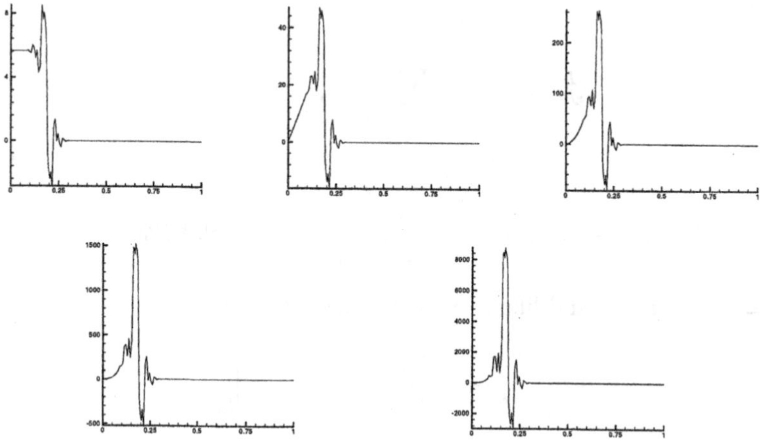

Fig. 1.8. Primal modified single scale functions $\xi^L_{5,k}$ near the left boundary for $d = 3$ and $\tilde{d} = 5$, $k = 1, \ldots, 5$.

Remark 1. One has for $x \in [0, 1]$

$$\begin{aligned} \xi^R_{j,2^j-\ell+d-\mu(d)-r}(1-x) &= \xi^L_{j,\ell-d+r}(x), \quad r = 0, \ldots, d-1, \\ \tilde{\xi}^R_{j,2^j-\tilde{\ell}+\tilde{d}-\mu(d)-r}(1-x) &= \tilde{\xi}^L_{j,\tilde{\ell}-\tilde{d}+r}(x), \quad r = 0, \ldots, \tilde{d}-1. \end{aligned} \tag{1.70}$$

and

$$\theta_{[j+1,m]}(x) = \theta_{[j+1,2^{j+1}-m-\mu(d)]}(1-x), \quad \theta = \xi, \tilde{\xi}. \qquad \square \tag{1.71}$$

Defining now

$$\check{\xi}_{j,k} := \begin{cases} \xi^X_{j,k}, & k \in I^X_j,\ X \in \{L, R\}, \\ \xi_{[j,k]}, & k \in I^0_j, \end{cases}$$

let

$$\check{\Xi}_j := \{\check{\xi}_{j,k} \ : \ k \in I_j\} \tag{1.72}$$

and similarly

$$\check{\tilde{\Xi}}_j := \{\tilde{\xi}^L_{j,k} \ : \ k \in \tilde{I}^L_j\} \cup \{\tilde{\xi}_{[j,k]} \ : \ k \in \tilde{I}^0_j\} \cup \{\tilde{\xi}^R_{j,k} \ : \ k \in \tilde{I}^R_j\}. \tag{1.73}$$

We use the notation $\check{\Xi}_j$ and $\check{\tilde{\Xi}}_j$ here since these systems in general are *not* biorthogonal, they will serve as intermediate systems of functions. Finally, define the multiresolution spaces (see (1.9) on p. 4) by

$$S_j := S(\check{\Xi}_j), \quad \tilde{S}_j := S(\check{\tilde{\Xi}}_j). \tag{1.74}$$

Proposition 1. *Under the above assumptions, we have*

1. *The spaces S_j and $\tilde{S}_j$ are nested, i.e.,*

$$S_j \subset S_{j+1}, \quad \tilde{S}_j \subset \tilde{S}_{j+1}, \quad j \geq j_0. \tag{1.75}$$

2. *The spaces S_j, $\tilde{S}_j$ are exact of order d, $\tilde{d}$, respectively, i.e.,*

$$\mathcal{P}_d([0,1]) \subset S_j, \quad \mathcal{P}_{\tilde{d}}([0,1]) \subset \tilde{S}_j, \quad j \geq j_0. \quad \square \tag{1.76}$$

The proof is straightforward. The first statement follows from the refinement relations whereas the second one is an immediate consequence of the construction of the boundary scaling functions. The above result shows that $\mathcal{S} = \{S_j\}_{j \geq j_0}$ and $\tilde{\mathcal{S}} = \{\tilde{S}_j\}_{j \geq j_0}$ in fact form a dual Multiresolution Analysis.

1.3.3 Biorthogonalization

By construction, the spanning sets $\check{\Xi}_j$ and $\check{\tilde{\Xi}}'_j$ from (1.72) and (1.73) have equal cardinality. It remains to verify next that these sets of functions are linearly independent and, which is a stronger property, that $\check{\Xi}_j$ and $\check{\tilde{\Xi}}'_j$ can be biorthogonalized. Thus, we seek coefficients $\boldsymbol{E}^X_j := (e^X_{j,k,m})_{k,m \in \tilde{I}^X_j}$, $\tilde{\boldsymbol{E}}^X_j := (\tilde{e}^X_{j,k,m})_{k,m \in \tilde{I}^X_j}$, $X \in \{L, R\}$, (recall $\tilde{I}^X_j \supseteq I^X_j$) such that the systems of functions

$$\Xi^X_j := \boldsymbol{E}^X_j \check{\Xi}_j, \qquad \tilde{\Xi}^X_j := \tilde{\boldsymbol{E}}^X_j \check{\tilde{\Xi}}_j, \tag{1.77}$$

satisfy

$$\left(\Xi^X_j, \tilde{\Xi}^X_j\right)_{0,(0,1)} = \left((\xi_{j,k}, \tilde{\xi}_{j,k'})_{0,(0,1)}\right)_{k,k' \in \tilde{I}^X_j} = \boldsymbol{I}_{\tilde{I}^X_j}, \quad X \in \{L, R\}. \tag{1.78}$$

Since (1.77) is a change of basis, the matrices $\boldsymbol{E}^X_j$, $\tilde{\boldsymbol{E}}^X_j$ have to be nonsingular. Defining the generalized gramian

$$\mathbf{\Gamma}_{j,X} := \left((\breve{\xi}_{j,k}, \breve{\tilde{\xi}}_{j,m})_{0,(0,1)} \right)_{k,m \in \tilde{I}_j^X}, \tag{1.79}$$

(1.78) is equivalent to

$$\boldsymbol{I}_X = \boldsymbol{E}_j^X \mathbf{\Gamma}_{j,X} (\tilde{\boldsymbol{E}}_j^X)^T,$$

i.e., $(\boldsymbol{E}_j^X)^{-1} (\tilde{\boldsymbol{E}}_j^X)^{-T} = \mathbf{\Gamma}_{j,X}$. Thus we can biorthogonalize $\breve{\Xi}_j$ and $\breve{\tilde{\Xi}}_j$ if and only if $\mathbf{\Gamma}_{j,X}$ is regular. This is confined by the following result. We will frequently use the notation $\boldsymbol{A}^{\updownarrow}$ to indicate the matrix that results from a matrix $\boldsymbol{A}$ by reversing the ordering of rows and columns (i.e., by 'counter-transposing').

Theorem 3. *The matrices $\mathbf{\Gamma}_{j,X}$ are always non-singular. Moreover, they have uniformly bounded condition numbers. That is, the matrices $\mathbf{\Gamma}_{j,L}$ are independent of j,*

$$\mathbf{\Gamma}_{j,L} = \mathbf{\Gamma}_L \tag{1.80}$$

while

$$\mathbf{\Gamma}_{j,R} = \mathbf{\Gamma}_L^{\updownarrow} \tag{1.81}$$

which means here $(\mathbf{\Gamma}_R)_{k,m} = (\mathbf{\Gamma}_L)_{2^j-\mu(d)-k,2^j-\mu(d)-m}$, $k, m \in \tilde{I}_j^R$.

Proof. By (1.66) we have for $r = 0, \ldots, d-1$ and $k = 0, \ldots, \tilde{d}-1$

$$\begin{aligned}
(\xi^L_{j,\ell-d+r}, \tilde{\xi}^L_{j,\tilde{\ell}-\tilde{d}+k})_{0,(0,1)} &= \sum_{\nu=-\ell_2+1}^{\ell-1} \sum_{\mu=-\tilde{\ell}_2+1}^{\tilde{\ell}-1} \tilde{\alpha}_{\nu,r} \alpha_{\mu,k} (\xi_{[j,\nu]}, \tilde{\xi}_{[j,\mu]})_{0,(0,1)} \\
&= \sum_{\nu=-\ell_2+1}^{\ell-1} \sum_{\mu=-\tilde{\ell}_2+1}^{\tilde{\ell}-1} \tilde{\alpha}_{\nu,r} \alpha_{\mu,k} \int_0^{2^j} \xi(x-\nu) \tilde{\xi}(x-\mu)\, dx.
\end{aligned}$$

Similarly one obtains for $r = d, \ldots, \tilde{d}-1$, $k = 0, \ldots, \tilde{d}-1$,

$$(\xi_{[j,\ell-d+r]}, \tilde{\xi}^L_{j,\tilde{\ell}-\tilde{d}+k})_{0,(0,1)} = \sum_{\mu=-\tilde{\ell}_2+1}^{\tilde{\ell}-1} \alpha_{\mu,k} \int_0^{2^j} \xi(x-(\ell-d+r)) \tilde{\xi}(x-\mu)\, dx.$$

Since for $j \geq j_0$ and $-\ell_2+1 \leq \nu \leq \ell-1, \quad -\tilde{\ell}_2+1 \leq \mu \leq \tilde{\ell}-1$

$$\int_0^{2^j} \xi(x-\nu)\tilde{\xi}(x-\mu)\, dx = \int_0^{\infty} \xi(x-\nu)\tilde{\xi}(x-\mu)\, dx$$

(1.80) follows, while (1.81) is an immediate consequence of the symmetry relations (1.70), (1.71).

Now suppose first that $d \geq 2$. Note that, by duality and the definition of $\xi^L_{j,k}$, $\tilde{\xi}^L_{j,k}$ (1.66), (1.67) and (1.60), we have

$$(\xi^L_{j,\ell-d+r}, \tilde{\xi}^L_{j,\tilde{\ell}-\tilde{d}+k})_{0,(0,1)} = \Big(\xi^L_{j,\ell-d+r}, \sum_{m=-\tilde{\ell}_2+1}^{\infty} \alpha_{m,k}\tilde{\xi}_{[j,m]}\Big)_{0,(0,1)}$$
$$= 2^{j/2}2^{kj}(\xi^L_{j,\ell-d+r}, (\cdot)^k)_{0,(0,1)},$$

for $k = 0, \ldots, \tilde{d}-1$. Thus

$$\mathbf{\Gamma}_L = \Big(2^{j/2}2^{kj}\,(\xi^L_{j,\ell-d+r}, (\cdot)^k)_{0,(0,1)}\Big)_{r,k=0}^{\tilde{d}-1} = \mathbf{\Gamma}_L(d,\tilde{d},\ell,0),$$

where we define more generally

$$\mathbf{\Gamma}_L(d,\tilde{d},\ell,\nu) := \Big(2^{j/2}2^{kj}\,(\xi^L_{j,\ell-d+r}, (\cdot)^{\nu+k})_{0,(0,1)}\Big)_{r,k=0}^{\tilde{d}-1}.$$

To show that $\mathbf{\Gamma}_L$ is nonsingular, it will be useful to keep track of the dependence of the various entities on the parameters $d, \tilde{d}, \ell, \tilde{\ell}$. Therefore, we write

$$\xi^L_{j,k}(x) = \xi^L_{j,k}(x \mid d, \tilde{d}, \ell), \qquad \tilde{\alpha}_{m,r} = \tilde{\alpha}_{m,r}(d,\tilde{d}) = \int_{\mathbb{R}} x^r\,{}_{d,\tilde{d}}\tilde{\xi}(x-m)dx.$$

Rewriting formula (3.4.11) in [52] in present terms (see also Theorem 6 on p. 84 and (2.2)) yields the relations

$$\frac{d}{dx}\xi^L_{j,\ell-d+r}(x \mid d,\tilde{d},\ell) = \begin{cases} -2^j\,{}_{d-1}\xi_{[j,\ell-\mu(d-1)]}(x), & r=0,\\ 2^j\big(r\xi^L_{j,\ell-d+r-\mu(d-1)}(x \mid d-1, \tilde{d}+1, \ell-\mu(d-1)) \\ \qquad -\tilde{\alpha}_{\ell-1,r}(d,\tilde{d})\,{}_{d-1}\xi_{[j,\ell-\mu(d-1)]}\big), & \\ & r=1,\ldots,d-1, \end{cases} \tag{1.82}$$

while for $k \geq \ell$

$$\frac{d}{dx}\,{}_d\xi_{[j,k]} = 2^j\left({}_{d-1}\xi_{[j,k+\mu(d-1)]} - {}_{d-1}\xi_{[j,k+1-\mu(d-1)]}\right). \tag{1.83}$$

These relations are obtained by straightforward calculations with the aid of

$$\tilde{\alpha}_{m,r}(d,\tilde{d}) - \tilde{\alpha}_{m-1,r}(d,\tilde{d}) = r\tilde{\alpha}_{m-\mu(d-1),r-1}(d-1,\tilde{d}+1),$$

for $r = 0, \ldots, d-1$, which in turn follow from the definition and the formula

$$\frac{d}{dx}\,{}_d\xi(x) = {}_{d-1}\xi(x+\mu(d-1)) - {}_{d-1}\xi(x-\mu(d)),$$

(see again (2.2) on p. 84). Therefore by (1.82) and (1.83), we have for any $k = 0, \ldots, \tilde{d}-1$,

$$\left(\xi^L_{j,\ell-d+r},(\cdot)^k\right)_{0,(0,1)} = -\frac{1}{k+1}\left(\frac{d}{dx}\xi^L_{j,\ell-d+r}(\cdot\mid d,\tilde d,\ell),(\cdot)^{k+1}\right)_{0,(0,1)}$$

$$= \begin{cases} \frac{2^j}{k+1}\left({}_{d-1}\xi_{[j,\ell-\mu(d-1)]},(\cdot)^{k+1}\right)_{0,(0,1)}, & \text{for } r=0,\\ \frac{2^j\tilde\alpha_{\ell-1,r}(d,\tilde d)}{k+1}\left({}_{d-1}\xi_{[j,\ell-\mu(d-1)]},(\cdot)^{k+1}\right)_{0,(0,1)} \\ \quad -\frac{2^j r}{k+1}\left(\xi^L_{j,\ell-(d-1)+r-1-\mu(d-1)}(\cdot\mid d-1,\tilde d+1,\ell-\mu(d-1)),(\cdot)^{k+1}\right)_{0,(0,1)} \\ & \text{for } r=1,\ldots,d-1,\\ \frac{2^j}{k+1}\left({}_{d-1}\xi_{[j,\ell-d+r-\mu(d-1)]}-{}_{d-1}\xi_{[j,\ell-d+r+1-\mu(d-1)]},(\cdot)^{k+1}\right)_{0,(0,1)}, \\ & \text{for } r=d,\ldots,\tilde d-1. \end{cases}$$

One readily concludes from the latter equation that $\mathbf{\Gamma}_L(d,\tilde d,\ell,0)$ is nonsingular if and only if $\mathbf{\Gamma}_L(d-1,\tilde d,\ell-\mu(d-1),1)$ is nonsingular. As mentioned before, we have set here

$${}_{d-1}\xi_{j,\ell-\mu(d)-(d-1)+r} = \begin{cases} \xi^L_{j,\ell-\mu(d-1)-(d-1)+r}(\cdot|d-1,\tilde d+1,\ell-\mu(d-1)), \\ & \text{for } r=0,\ldots,d-2,\\ {}_{d-1}\xi_{[j,\ell-\mu(d)-(d-1)+r]}, \\ & \text{for } r=d-1,\ldots,\tilde d-1. \end{cases}$$

Repeating this argument provides

$$\det\mathbf{\Gamma}_L\neq 0 \quad \text{iff} \quad \det\mathbf{\Gamma}_L(1,\tilde d,\hat\ell,d-1)\neq 0, \tag{1.84}$$

where $\hat\ell := \ell-\mu(d-1)-\cdots-\mu(1)$. Thus we are finished as soon as we have shown that for any $\ell,\tilde d\in\mathbb{N}$, $\nu\in\mathbb{N}\cup\{0\}$ the matrix $\mathbf{\Gamma}_L(1,\tilde d,\ell,\nu)$ is nonsingular.

To this end, note that for $d=1$, i.e., $\xi(x)=\chi_{[0,1)}$, we have $\ell_1=0$, $\ell_2=1$. Thus (1.66) gives

$$\xi^L_{j,\ell-1}(x) = \sum_{m=0}^{\ell-1}\tilde\alpha_{m,0}\xi_{[j,m]}(x) = \sum_{m=0}^{\ell-1}\xi_{[j,m]}(x) \tag{1.85}$$

because, by the normalization, $\tilde\alpha_{m,0}=\int_{\mathbb{R}}\tilde\xi(x-m)\,dx=1$. Therefore, one has

$$\begin{aligned} 2^{j(k+1/2)}\left(\xi^L_{j,\ell-1},(\cdot)^k\right)_{0,(0,1)} &= 2^{j(k+1/2)}\sum_{m=0}^{\ell-1}\int_0^1 2^{j/2}\chi_{[2^{-j}m,2^{-j}(m+1))}(x)x^k\,dx \\ &= 2^{j(k+1)}\int_0^{2^{-j}\ell} x^k\,dx, \end{aligned}$$

i.e., for $k\geq 0$,

$$2^{j(k+1/2)}\left(\xi^L_{j,\ell-1},(\cdot)^k\right)_{0,(0,1)} = \frac{\ell^{k+1}}{k+1}. \tag{1.86}$$

Moreover, we have for $\nu = \ell, \ldots, \tilde{\ell} - 1$

$$2^{j(k+1/2)} \left(\xi_{[j,\nu]}, (\cdot)^k\right)_{0,(0,1)} = 2^{j(k+1)} \int_{2^{-j}\nu}^{2^{-j}(\nu+1)} x^k dx = \frac{1}{k+1}((\nu+1)^{k+1} - \nu^{k+1}).$$

Thus $\mathbf{\Gamma}_L(1, \tilde{d}, \ell, \nu)$ takes the form

$$\mathbf{\Gamma}_L(1, \tilde{d}, \ell, \nu) = \begin{pmatrix} \frac{\ell^{\nu+1}}{\nu+1} & \frac{\ell^{\nu+2}}{\nu+2} & \cdots & \frac{\ell^{\nu+\tilde{d}}}{\nu+\tilde{d}} \\ \frac{(\ell+1)^{\nu+1}-\ell^{\nu+1}}{\nu+1} & \frac{(\ell+1)^{\nu+2}-\ell^{\nu+2}}{\nu+2} & \cdots & \frac{(\ell+1)^{\nu+\tilde{d}}-\ell^{\nu+\tilde{d}}}{\nu+\tilde{d}} \\ \vdots & \vdots & & \vdots \\ \frac{\tilde{\ell}^{\nu+1}-(\tilde{\ell}-1)^{\nu+1}}{\nu+1} & \frac{\tilde{\ell}^{\nu+2}-(\tilde{\ell}-1)^{\nu+2}}{\nu+2} & \cdots & \frac{\tilde{\ell}^{\nu+\tilde{d}}-(\tilde{\ell}-1)^{\nu+\tilde{d}}}{\nu+\tilde{d}} \end{pmatrix}.$$

Adding the first row to the second one, adding the result to the third row and so on, produces the matrix

$$\begin{pmatrix} \frac{\ell^{\nu+1}}{\nu+1} & \frac{\ell^{\nu+2}}{\nu+2} & \cdots & \frac{\ell^{\nu+\tilde{d}}}{\nu+\tilde{d}} \\ \frac{(\ell+1)^{\nu+1}}{\nu+1} & \frac{(\ell+1)^{\nu+2}}{\nu+2} & \cdots & \frac{(\ell+1)^{\nu+\tilde{d}}}{\nu+\tilde{d}} \\ \vdots & \vdots & & \vdots \\ \frac{\tilde{\ell}^{\nu+1}}{\nu+1} & \frac{\tilde{\ell}^{\nu+2}}{\nu+2} & \cdots & \frac{\tilde{\ell}^{\nu+\tilde{d}}}{\nu+\tilde{d}} \end{pmatrix}.$$

Dividing the ith row by $(\ell + i - 1)^{\nu+1}$ and multiplying then the i-th column of the resulting matrix by $\nu + i$ finally produces a Vandermonde matrix which is nonsingular. Taking $\nu = 0$, this confirms the claim for $d = 1$. By (1.84), the case $\nu = d - 1$ (with ℓ replaced by $\hat{\ell}$) verifies the assertion for any $d \geq 2$ which completes the proof. □

Using similar arguments, the result can also be obtained for orthogonal scaling functions and for those biorthogonal scaling functions that are produced by Theorem 6 below when starting with orthogonal functions. However, a corresponding result is still not known for general scaling systems.

The dual functions $\tilde{\xi}_{5,k}$ arising from those shown in Fig. 1.8 by choosing $\boldsymbol{E}_j = \boldsymbol{I}_{I_j}$ are shown in Fig. 1.9. The above findings can be summarized as follows.

Corollary 1. *Under the above assumptions, the following holds:*

1. *The collections Ξ_j, $\tilde{\Xi}_j$ defined by (1.77) are biorthogonal.*
2. *For the cardinalities, we obtain*

$$\dim S_j = \dim \tilde{S}_j = \#I_j = 2d + 2^j - \mu(d) - 2\ell + 1 = 2^j - 2(\ell - d) - \mu(d) + 1. \tag{1.87}$$

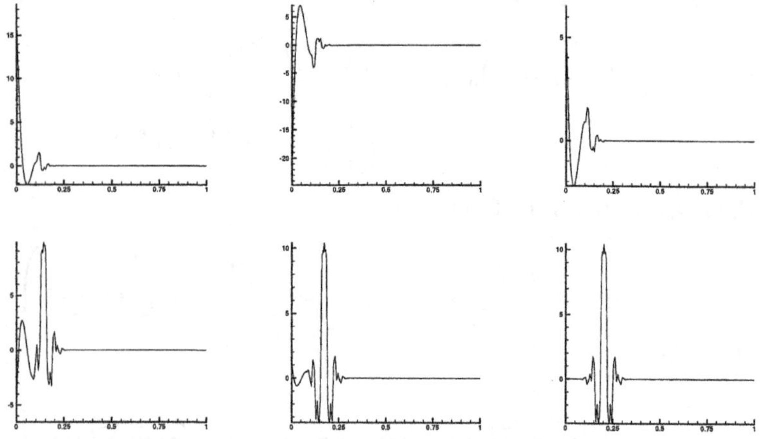

Fig. 1.9. Dual scaling functions $\tilde{\xi}_{5,k}$, $k = 1, \ldots, 6$ for $d = 3$, $\tilde{d} = 5$ using $\boldsymbol{E}_j = \boldsymbol{I}_{I_j}$ in the biorthogonalization.

3. *The support of the basis functions scale properly, i.e.,*

$$\operatorname{diam}(\operatorname{supp} \xi_{j,k}), \ \operatorname{diam}(\operatorname{supp} \tilde{\xi}_{j,k}) \sim 2^j, \quad j \geq j_0. \tag{1.88}$$

4. *The bases* $\{\Xi_j\}_{j \geq j_0}$, $\{\tilde{\Xi}_j\}_{j \geq j_0}$ *are uniformly stable.*
5. *The projectors*

$$Q_j v := (v, \tilde{\Xi}_j)_{0,(0,1)} \, \Xi_j, \quad Q_j^* v := (v, \Xi_j)_{0,(0,1)} \, \tilde{\Xi}_j \tag{1.89}$$

are uniformly bounded.

6. *The spaces* $S_j = S(\Xi_j)$ *and* $\tilde{S}_j = S(\tilde{\Xi}_j)$ *are nested and exact of order* d *and* $\tilde{d}$ *respectively.*
7. *Let the* Q_j *be defined by* (1.89). *Then one has (with* $Q_{j_0-1} := 0$*)*

$$\Big(\sum_{j=j_0}^{\infty} 2^{2sj} \, \|(Q_j - Q_{j-1})v\|_{0,(0,1)}^2 \Big)^{1/2} \sim \|v\|_{s,(0,1)}, \tag{1.90}$$

where $s \in (-\tilde{\gamma}, \gamma)$ *for* $\gamma = d - 1/2$ *and* $\tilde{\gamma} := \sup\{s : \tilde{\xi} \in H^s(\mathbb{R})\}$. □

1.3.4 Refinement Matrices

Now we identify the refinement matrices corresponding to Ξ_j and $\tilde{\Xi}_j$. From Lemma 1 on p. 19 we infer that $\breve{\Xi}_j$ satisfies the refinement equation with

$$\check{\boldsymbol{M}}_{j,0} := \tag{1.91}$$

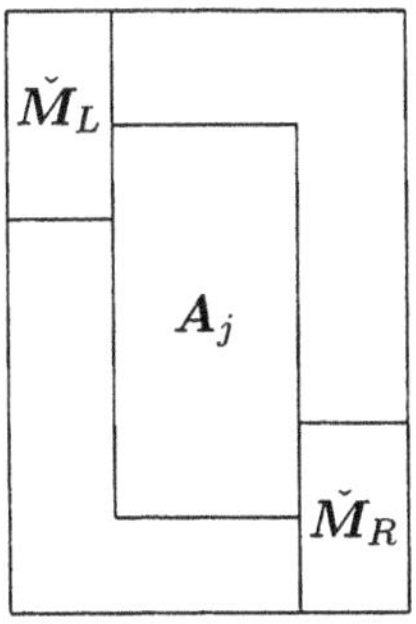

where $\check{\boldsymbol{M}}_L$, $\check{\boldsymbol{M}}_R$ are $(d+\ell+\ell_2-1)\times d$ blocks of the form

$$(\check{\boldsymbol{M}}_L)_{m,k} = \begin{cases} 2^{-(k-\ell+d+1/2)}\,\delta_{k,m}, & m\in\{\ell-d,\dots,\ell-1\} = I_j^L,\\ 2^{-(k-\ell+d+1/2)}\,\tilde{\alpha}_{m,k-\ell+d}, & m=\ell,\dots,2\ell+\ell_1-1,\\ \tilde{\beta}^L_{j,m,k-\ell+d}, & m=2\ell+\ell_1,\dots,\ell_2+2\ell-2,\end{cases} \tag{1.92}$$

for $k\in I_j^L$ and $\check{\boldsymbol{M}}_R = \check{\boldsymbol{M}}_L^{\updownarrow}$, i.e.,

$$(\check{\boldsymbol{M}}_R)_{2^j-\mu(d)-m,2^j-\mu(d)-k} = (\check{\boldsymbol{M}}_L)_{m,k},\quad m=\ell-d,\dots,\ell_2+2\ell-2,\ k\in I_j^L.$$

Moreover, $\boldsymbol{A}_j$ has the form

$$(\boldsymbol{A}_j)_{m,k} = \frac{1}{\sqrt{2}}\,a_{m-2k},\qquad 2\ell+\ell_1\le m\le \ell_2+2^{j+1}-2(\ell+\mu(d)),\quad k\in I_j^0. \tag{1.93}$$

The structure of the refinement matrix $\check{\tilde{\boldsymbol{M}}}_{j,0}$ corresponding to $\check{\tilde{\Xi}}_j$ defined in (1.73) on p. 25 is completely analogous and results from replacing ℓ,ℓ_1,ℓ_2,d by $\tilde{\ell},\tilde{\ell}_1,\tilde{\ell}_2,\tilde{d}$, respectively, i.e.,

$$\check{\tilde{\boldsymbol{M}}}_{j,0} = \tag{1.94}$$

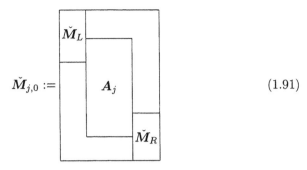

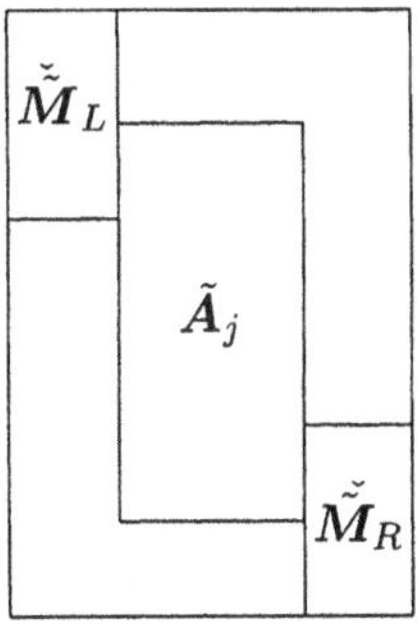

with the $(\tilde{d}+\tilde{\ell}+\tilde{\ell}_2+1)\times\tilde{d}$ blocks

$$(\check{\tilde{\boldsymbol{M}}}_L)_{m,k} = \begin{cases} 2^{-(k-\tilde{\ell}+\tilde{d}+1/2)}\,\delta_{k,m}, & m\in\{\tilde{\ell}-\tilde{d},\dots,\tilde{\ell}-1\} = \tilde{I}_j^L,\\ 2^{-(k-\tilde{\ell}+\tilde{d}+1/2)}\,\alpha_{m,k-\tilde{\ell}+\tilde{d}}, & m=\tilde{\ell},\dots,2\tilde{\ell}+\tilde{\ell}_1-1,\\ \beta^L_{j,m,k-\tilde{\ell}+\tilde{d}}, & m=2\tilde{\ell}+\tilde{\ell}_1,\dots,\tilde{\ell}_2+2\tilde{\ell}-2,\end{cases} \tag{1.95}$$

for $k \in \tilde{I}_j^L$ and

$$\check{\tilde{\boldsymbol{M}}}_R = \left(\check{\tilde{\boldsymbol{M}}}_L\right)^{\updownarrow} \tag{1.96}$$

as well as

$$(\tilde{\boldsymbol{A}}_j)_{m,k} = \frac{1}{\sqrt{2}}\tilde{a}_{m-2k}, \qquad 2\tilde{\ell}+\tilde{\ell}_1 \le m \le \tilde{\ell}_2 + 2^{j+1} - 2(\tilde{\ell}+\mu(d)), \quad k \in \tilde{I}_j^0. \tag{1.97}$$

The refinement matrices for the biorthogonal systems Ξ_j and $\tilde{\Xi}_j$ in (1.77) can easily be determined by

$$\Xi_j = \boldsymbol{E}_j \check{\Xi}_j = \boldsymbol{E}_j \check{\boldsymbol{M}}_{j,0}^T \check{\Xi}_{j+1} = \boldsymbol{E}_j \check{\boldsymbol{M}}_{j,0}^T \boldsymbol{E}_{j+1}^{-1} \Xi_{j+1},$$

where

$$\boldsymbol{E}_j := \begin{pmatrix} \boldsymbol{\Gamma}_{j,L} & 0 & 0 \\ 0 & \boldsymbol{I}_{\tilde{I}_j^0} & 0 \\ 0 & 0 & \boldsymbol{\Gamma}_{j,R} \end{pmatrix}, \tag{1.98}$$

i.e.,

$$\boldsymbol{M}_{j,0} := \boldsymbol{E}_{j+1}^{-T} \check{\boldsymbol{M}}_{j,0} \boldsymbol{E}_j^T$$

and similarly $\tilde{\boldsymbol{M}}_{j,0} := \tilde{\boldsymbol{E}}_{j+1}^{-T} \check{\tilde{\boldsymbol{M}}}_{j,0} \tilde{\boldsymbol{E}}_j^T$.

A particular choice of interest is $\boldsymbol{E}_j = \boldsymbol{I}_{I_j}$ and then $\tilde{\boldsymbol{E}}_j = \tilde{C}_j^T$, where

$$\tilde{\boldsymbol{C}}_j := \begin{pmatrix} \boldsymbol{\Gamma}_L^{-1} & 0 & 0 \\ 0 & \boldsymbol{I}_{\tilde{I}_j^0} & 0 \\ 0 & 0 & \boldsymbol{\Gamma}_R^{-1} \end{pmatrix}, \tag{1.99}$$

Then, we obtain

$$\tilde{\boldsymbol{M}}_{j,0} = \tilde{\boldsymbol{C}}_{j+1}^{-1} \check{\tilde{\boldsymbol{M}}}_{j,0} \tilde{\boldsymbol{C}}_j. \tag{1.100}$$

Keeping (1.94) in mind and splitting $\check{\tilde{\boldsymbol{M}}}_R$ into two blocks such as

$$\check{\tilde{\boldsymbol{M}}}_L = \begin{bmatrix} \boldsymbol{D} \\ \boldsymbol{K} \end{bmatrix}, \quad \boldsymbol{D} = 2^{-(k-\tilde{\ell}+\tilde{d}+1/2)}\delta_{k,m}, \quad k,m \in \tilde{I}_j^L,$$

with $\boldsymbol{K}$ defined by (1.95) one easily confirms that

$$\tilde{\boldsymbol{M}}_{j,0} = \tag{1.101}$$

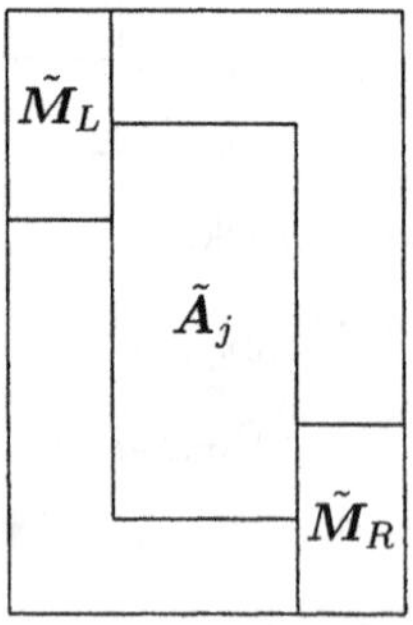

where now

$$\tilde{M}_L = \Gamma_L D \begin{bmatrix} \Gamma_L^{-1} \\ K \end{bmatrix} \Gamma_L^{-1}, \quad \tilde{M}_R = \tilde{M}_L^{\updownarrow}, \tag{1.102}$$

and $\tilde{A}_j$ remains the same as in (1.97).

1.3.5 Biorthogonal Wavelets on $(0,1)$

So far, we have constructed a collection Ξ_j, $\tilde{\Xi}_j$ of biorthogonal single scale basis functions. Our next goal is to determine the corresponding collections Υ_j, $\tilde{\Upsilon}_j$ of biorthogonal wavelets. In the following, we restrict ourselves to the case $E_j = I_{I_j}$. However, note that all other cases can easily be obtained from this.

Stable Completion. We follow a general construction principle called *stable completion*, see Sec. 1.1.7 and also [25]. The idea is to determine first *some* initial complement (*initial stable completion*) of $S(\Xi_j)$ in $S(\Xi_{j+1})$. In the second step, we project this complement onto the desired complement $(Q_{j+1} - Q_j)S(\Xi_{j+1})$. Given Ξ_j and $M_{j,0}$, any $M_{j,1} \in [\ell_2(J_j), \ell_2(I_{j+1})]$ is called *(uniformly) stable completion* of $M_{j,0}$ if

$$\|M_j\|, \|M_j^{-1}\| = \mathcal{O}(1), \qquad j \to \infty,$$

where $M_j := (M_{j,0}, M_{j,1})$ Then, the following result describes the construction principle.

Proposition 2. *[25] Let $\{\Xi_j\}$, $\{\tilde{\Xi}_j\}$, $M_{j,0}$ and $\tilde{M}_{j,0}$ be related as above. Suppose that $\check{M}_{j,1}$ is some stable completion of $M_{j,0}$ and that $\check{G}_j = \check{M}_j^{-1}$. Then*

$$M_{j,1} := (I - M_{j,0}\tilde{M}_{j,0}^T)\check{M}_{j,1} \tag{1.103}$$

is also a stable completion and $G_j = M_j^{-1}$ has the form

$$G_j = \begin{bmatrix} \tilde{M}_{j,0}^T \\ \check{G}_{j,1} \end{bmatrix}. \tag{1.104}$$

Moreover, the collections

$$\Xi_j := M_{j,1}^T \Xi_{j+1}, \quad \tilde{\Upsilon}_j := \check{G}_{j,1}, \tilde{\Xi}_{j+1} \tag{1.105}$$

form biorthogonal systems,

$$\langle \Xi_j, \tilde{\Upsilon}_j \rangle = I, \quad \langle \Upsilon_j, \tilde{\Xi}_j \rangle = \langle \Xi_j, \tilde{\Upsilon}_j \rangle = 0, \tag{1.106}$$

so that

$$(Q_{j+1} - Q_j)S(\Xi_{j+1}) = S(\Upsilon_j), \quad (Q_{j+1}^* - Q_j^*)S(\tilde{\Xi}_{j+1}) = S(\tilde{\Upsilon}_j). \qquad \square \tag{1.107}$$

Gauß-type Elimination of Refinement Matrices. In order to obtain an initial stable completion, we perform a Gauß-type elimination for the refinement matrix $\boldsymbol{M}_{j,0} = \check{\boldsymbol{M}}_{j,0}$. It can be shown [54] that this process is well defined for biorthogonal B-spline scaling functions. Since the proof makes heavy use of special properties of B-splines it is still an open question how to construct wavelet bases on the interval for general scaling systems in order to preserve all relevant properties (see also the remarks concerning Theorem 3 on p. 26 above).

As a first step we will construct certain stable complement bases for the spaces S_j corresponding to a stable completion of the refinement matrices of Ξ_j in the sense of Proposition 2. In a second step these initial complements will be projected into the desired ones employing again the above tools.

The construction of the initial stable completion $\check{\boldsymbol{M}}_{j,1}$ of $\boldsymbol{M}_{j,0}$ in (1.91) on p. 31 consists of several steps, each of which involves different matrices which are described most conveniently in a schematic block form. All these matrices will depend only *weakly* on the scale j which means that the entries of the various blocks remain the same and only the size of the central blocks depends on j. To describe the size of the involved blocks accurately it will be convenient to abbreviate

$$\begin{aligned} p = p(j) &:= \#I_j^0 = 2^j - 2\ell - \mu(d) + 1, \\ q = q(j) &:= 2p + d - 1 = 2^{j+1} - 4\ell - 2\mu(d) + d + 1 \end{aligned}$$

and keep in mind that $\ell_1 = -\left\lfloor \frac{d}{2} \right\rfloor$, $\ell_2 = \left\lceil \frac{d}{2} \right\rceil$.

In these terms the interior block $\boldsymbol{A}_j$ in (1.91) on p. 31 is a $q \times p$ matrix of the form

$$\boldsymbol{A}_j = \frac{1}{\sqrt{2}} \begin{pmatrix} a_{\ell_1} & 0 & \cdots & 0 \\ a_{\ell_1+1} & 0 & & \vdots \\ a_{\ell_1+2} & a_{\ell_1} & & \\ \vdots & \vdots & & \vdots \\ a_{\ell_2} & a_{\ell_2-2} & & \\ 0 & a_{\ell_2-1} & & 0 \\ 0 & a_{\ell_2} & & a_{\ell_1} \\ \vdots & & & \vdots \\ & & a_{\ell_2} & a_{\ell_2-2} \\ & & 0 & a_{\ell_2-1} \\ & & 0 & a_{\ell_2} \end{pmatrix} \qquad (1.108)$$

where $a = \{a_k\}_{k=\ell_1}^{\ell_2}$ is the mask of $\xi = {}_d\xi$, (1.41) on p. 15.

The core ingredient of our construction is a factorization of $\boldsymbol{A}_j$ and later of $\boldsymbol{M}_{j,0}$ which is inspired by similar considerations for the biinfinite case in [55]. Employing suitable Gauß-type eliminations we will successively reduce upper and lower bands from $\boldsymbol{A}_j$. Suppose that after i steps the resulting matrix $\boldsymbol{A}_j^{(i)}$ has the form

$$
\boldsymbol{A}_j^{(i)} = \begin{pmatrix}
0 & 0 & & 0 \\
\vdots & \vdots & \Big\}\left\lceil \frac{i}{2} \right\rceil & \vdots \\
0 & 0 & & \\
a^{(i)}_{\ell_1+\lceil \frac{i}{2} \rceil} & 0 & & \\
a^{(i)}_{\ell_1+\lceil \frac{i}{2} \rceil+1} & 0 & & \\
\vdots & a^{(i)}_{\ell_1+\lceil \frac{i}{2} \rceil} & & \\
\vdots & \vdots & & \\
a^{(i)}_{\ell_2-\lfloor \frac{i}{2} \rfloor} & & & \\
0 & & & \\
\vdots & & & \\
 & & & a^{(i)}_{\ell_2-\lfloor \frac{i}{2} \rfloor} \\
 & & & 0 \\
 & & \left\lfloor \frac{i}{2} \right\rfloor \Big\{ & \vdots \\
0 & & & 0
\end{pmatrix}, \quad \boldsymbol{A}_j^{(0)} := \boldsymbol{A}_j. \tag{1.109}
$$

Defining

$$
\boldsymbol{U}_{i+1} := \begin{bmatrix} 1 & -\dfrac{a^{(i)}_{\ell_1+\lceil \frac{i}{2} \rceil}}{a^{(i)}_{\ell_1+\lceil \frac{i}{2} \rceil+1}} \\ 0 & 1 \end{bmatrix}, \quad \boldsymbol{L}_{i+1} := \begin{bmatrix} 1 & 0 \\ -\dfrac{a^{(i)}_{\ell_2-\lfloor \frac{i}{2} \rfloor}}{a^{(i)}_{\ell_2-\lfloor \frac{i}{2} \rfloor-1}} & 1 \end{bmatrix}, \tag{1.110}
$$

and setting

$$
\begin{aligned}
\boldsymbol{H}_j^{(2i-1)} &:= \operatorname{diag}\Big(\boldsymbol{I}^{(i-1)}, \underbrace{\boldsymbol{U}_{2i-1}, \ldots, \boldsymbol{U}_{2i-1}}_{p}, \boldsymbol{I}^{(d-i)}\Big) \in \mathbb{R}^{q \times q} \\
\boldsymbol{H}_j^{(2i)} &:= \operatorname{diag}\Big(\boldsymbol{I}^{(d-i)}, \underbrace{\boldsymbol{L}_{2i}, \ldots, \boldsymbol{L}_{2i}}_{p}, \boldsymbol{I}^{(i-1)}\Big) \in \mathbb{R}^{q \times q}
\end{aligned} \tag{1.111}
$$

one easily confirms that indeed

$$
\boldsymbol{A}_j^{(i)} = \boldsymbol{H}_j^{(i)} \boldsymbol{A}_j^{(i-1)}, \tag{1.112}
$$

provided that $\boldsymbol{H}_j^{(i)}$ is well-defined, which in turns means that $a^{(i-1)}_{\ell_1+\lceil \frac{i-1}{2} \rceil+1}$, $a^{(i-1)}_{\ell_2-\lfloor \frac{i-1}{2} \rfloor-1}$ have to be different from zero whenever $a^{(i-1)}_{\ell_1+\lceil \frac{i-1}{2} \rceil}$, $a^{(i-1)}_{\ell_2-\lfloor \frac{i-1}{2} \rfloor}$ are different from zero. For $i = 1$ this is clearly the case. More generally, the following result holds.

Remark 2. For cardinal B-splines, one has

$$a^{(i)}_{\ell_1+\lceil\frac{i}{2}\rceil},\ldots,a^{(i)}_{\ell_2-\lfloor\frac{i}{2}\rfloor}\neq 0. \tag{1.113}$$

Note that this result certainly not holds for arbitrary scaling functions. □

Hence, $\boldsymbol{H}_j^{(i)}$ is well-defined for $i=1,\ldots,d$, and $\boldsymbol{A}_j^{(d)}$ has the form

$$\boldsymbol{A}_j^{(d)}=\begin{pmatrix} \left.\begin{matrix}0 & 0\\ \vdots & \vdots\\ 0 & 0\end{matrix}\right\}\ell_2=\lceil\frac{d}{2}\rceil & & \\ \begin{matrix} b & 0\\ 0 & 0\\ 0 & b\\ \vdots & 0\end{matrix} & \ddots & \\ & & b\\ -\ell_1=\lfloor\frac{d}{2}\rfloor & \left\{\begin{matrix}0\\ \vdots\\ 0\end{matrix}\right. & \end{pmatrix}, \tag{1.114}$$

where

$$b:=a^{(d)}_{\ell_1+\lceil\frac{d}{2}\rceil}=a^{(d)}_{\ell_1+\ell_2}=a^{(d)}_{\mu(d)}\neq 0. \tag{1.115}$$

Obviously, $\boldsymbol{A}_j^{(d)}$ has full rank p and

$$\boldsymbol{B}_j:=\begin{pmatrix} 0 & \cdots & 0 & b^{-1} & 0 & 0 & 0 & \cdots & & & & \\ \underbrace{0 & \cdots & 0}_{\lceil\frac{d}{2}\rceil=\ell_2} & 0 & 0 & b^{-1} & 0 & \cdots & & & & \\ & & & & & & & \ddots & & & & \\ & & & & & & & & b^{-1} & \overbrace{0 & \cdots & 0}^{\lfloor\frac{d}{2}\rfloor=-\ell_1} \end{pmatrix} \tag{1.116}$$

satisfies

$$\boldsymbol{B}_j\boldsymbol{A}_j^{(d)}=\boldsymbol{I}^{(p)}. \tag{1.117}$$

Similarly, defining

$$\boldsymbol{F}_j=\begin{pmatrix} \left.\begin{matrix}0 & 0\\ \vdots & \vdots\\ 0 & 0\end{matrix}\right\}\ell_2-1 & & \\ \begin{matrix} 1 & 0\\ 0 & 0\\ 0 & 1\\ \vdots & 0\end{matrix} & \ddots & \\ & & 1\\ -\ell_1+1 & \left\{\begin{matrix}0\\ \vdots\\ 0\end{matrix}\right. & \end{pmatrix}\in\mathbb{R}^{q\times p} \tag{1.118}$$

essentially by shifting up each row of B_j^T by one, we have

$$B_j F_j = 0. \tag{1.119}$$

After these preparations we have to pad the matrices $A_j^{(d)}$, B_j, F_j according to (1.91) on p. 31 to form matrices of the right size. The corresponding expanded versions will be denoted by $\hat{A}_j^{(d)}$, $\hat{B}_j$, $\hat{F}_j$, respectively. To this end, let

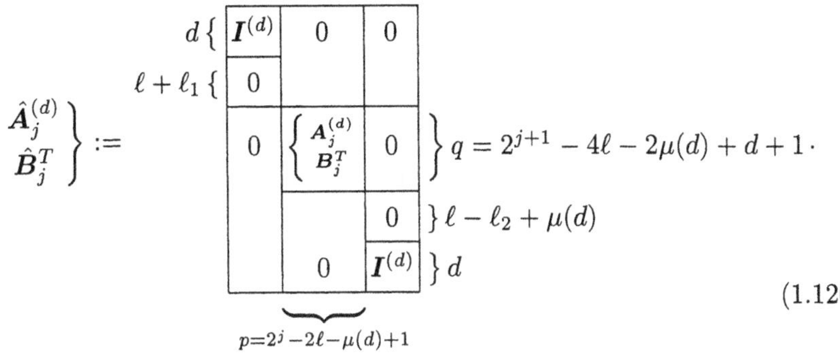

$$
\left.\begin{matrix}\hat{A}_j^{(d)}\\ \hat{B}_j^T\end{matrix}\right\} :=
\begin{array}{c}
d\{ \\
\ell+\ell_1\{ \\
\\
\\
\\
\end{array}
\begin{array}{|c|c|c|}
\hline
I^{(d)} & 0 & 0 \\
\hline
0 & & \\
\hline
& 0\ \begin{array}{|c|c|}\hline A_j^{(d)} & 0 \\ B_j^T & \\\hline\end{array} & \\
\hline
& & 0\ \}\ell-\ell_2+\mu(d) \\
\hline
& 0 & I^{(d)}\ \}d \\
\hline
\end{array}
\left.\begin{matrix}\\ \\ \}\ q=2^{j+1}-4\ell-2\mu(d)+d+1\cdot\\ \\ \\ \end{matrix}\right.
$$

$$\underbrace{\qquad\qquad}_{p=2^j-2\ell-\mu(d)+1}$$

$$\tag{1.120}$$

In fact, recalling (1.55) on p. 21 and noting that

$$d+\ell+\ell_1 = \ell+\ell_2 = \ell-\ell_2+\mu(d)+d,$$

one readily confirms that $\hat{A}_j^{(d)}$, $\hat{B}_j^T$ are $(\#I_{j+1}) \times (\#I_j)$ matrices.
Note that always

$$\#I_{j+1} - \#I_j = 2^j$$

is valid independent of $\ell, \tilde{\ell}, d, \tilde{d}$. Thus, a completion of $\hat{A}_j^{(d)}$ has to be a $(\#I_{j+1}) \times 2^j$ matrix. To this end, consider

$$\hat{F}_j :=\qquad\qquad\qquad\qquad\qquad ,\qquad (1.121)$$

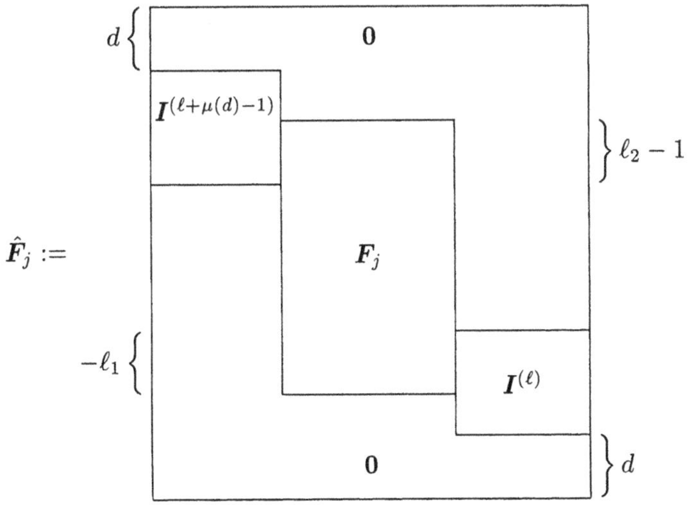

which is a $(\#I_{j+1}) \times 2^j$ matrix.

Lemma 2. *The following relations hold:*

$$\hat{B}_j \hat{A}_j^{(d)} = I^{(\#I_j)}, \quad \hat{F}_j^T \hat{F}_j = I^{(2^j)}, \tag{1.122}$$

and

$$\hat{B}_j \hat{F}_j = 0, \quad \hat{F}_j^T \hat{A}_j^{(d)} = 0. \quad \square \tag{1.123}$$

The factorization of A_j induced by (1.112) is easily carried over to a factorization of $\hat{A}_j$ which is defined by (1.120) with $A_j^{(d)}$ replaced by A_j given by (1.108). In fact, let

$$\hat{H}_j^{(i)} := \operatorname{diag}\left(I^{(\ell+\ell_2)}, H_j^{(i)}, I^{(\ell+\ell_2)}\right) \tag{1.124}$$

denote the corresponding expansions of the elimination matrices $H_j^{(i)}$. One easily checks that the above block structure leads to the following relations.

Lemma 3. *The matrix $\hat{A}_j$ can be factorized as*

$$\hat{A}_j = \hat{H}_j^{-1} \hat{A}_j^{(d)} \tag{1.125}$$

where

$$\hat{H}_j^{-1} = \left(\hat{H}_j^{(1)}\right)^{-1} \cdots \left(\hat{H}_j^{(d)}\right)^{-1} \tag{1.126}$$

and

$$\left(\hat{H}_j^{(2i-1)}\right)^{-1} = \operatorname{diag}\left(I^{(\ell+\ell_2+i-1)}, \underbrace{U_{2i-1}^{-1}, \ldots, U_{2i-1}^{-1}}_{p}, I^{(\ell+\ell_2+d-i)}\right),$$

$$\tag{1.127}$$

$$\left(\hat{H}_j^{(2i)}\right)^{-1} = \operatorname{diag}\left(I^{(\ell+\ell_2+d-i)}, \underbrace{L_{2i}^{-1}, \ldots, L_{2i}^{-1}}_{p}, I^{(\ell+\ell_2+i-1)}\right). \quad \square$$

As in the biinfinite case one has that the matrix $\hat{H}_j^{-1}$ is d-banded. The relevance of the factorization (1.125) relies on the observation that

$$M_{j,0} = P_j \hat{A}_j = P_j \hat{H}_j^{-1} \hat{A}_j^{(d)} \tag{1.128}$$

where

$$P_j = \begin{pmatrix} M_L & & \\ & I^{(\#I_{j+1}-2d)} & \\ & & M_R \end{pmatrix} \in \mathbb{R}^{(\#I_{j+1}) \times (\#I_{j+1})}. \tag{1.129}$$

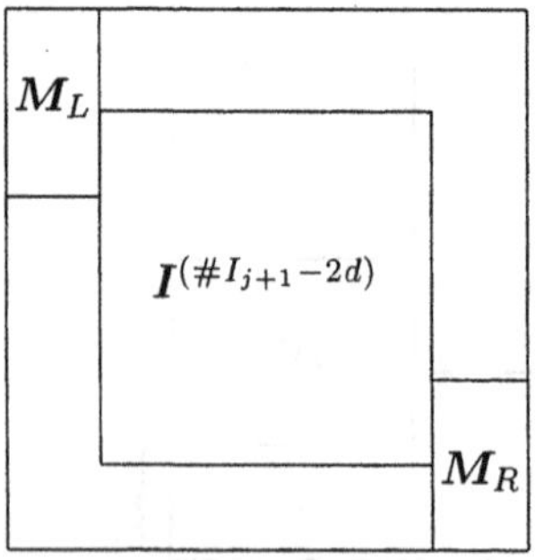

We are now in a position to state the main result of this section.

Proposition 3. *The matrices*

$$\check{\boldsymbol{M}}_{j,1} := \boldsymbol{P}_j \hat{\boldsymbol{H}}_j^{-1} \hat{\boldsymbol{F}}_j \tag{1.130}$$

are uniformly stable completions of the refinement matrices $\boldsymbol{M}_{j,0}$ (1.91) *on p. 31 for the bases* Ξ_j. *Moreover, the inverse*

$$\check{\boldsymbol{G}}_j = \begin{bmatrix} \check{\boldsymbol{G}}_{j,0} \\ \check{\boldsymbol{G}}_{j,1} \end{bmatrix}$$

of $\check{\boldsymbol{M}}_j = \left(\boldsymbol{M}_{j,0}, \check{\boldsymbol{M}}_{j,1}\right)$ *is given by*

$$\check{\boldsymbol{G}}_{j,0} = \hat{\boldsymbol{B}}_j \hat{\boldsymbol{H}}_j \boldsymbol{P}_j^{-1}, \quad \check{\boldsymbol{G}}_{j,1} = \hat{\boldsymbol{F}}_j^T \hat{\boldsymbol{H}}_j \boldsymbol{P}_j^{-1}. \quad \Box \tag{1.131}$$

Biorthogonal Wavelet Bases. It merely remains to put all the collected ingredients together to formulate the following result.

Theorem 4. *Adhering to the above notation for* $\boldsymbol{M}_{j,0}$, $\tilde{\boldsymbol{M}}_{j,0}$, $\check{\boldsymbol{M}}_{j,1}$ *defined by* (1.91), (1.101) *and* (1.130) *on pp. 31, 32 and 39, respectively, let*

$$\boldsymbol{M}_{j,1} := \left(\boldsymbol{I}^{(\# I_{j+1})} - \boldsymbol{M}_{j,0} \tilde{\boldsymbol{M}}_{j,0}^T\right) \check{\boldsymbol{M}}_{j,1}. \tag{1.132}$$

Then the following statements hold:

1. *The* $\boldsymbol{M}_{j,1}$ *are uniformly stable completions of the* $\boldsymbol{M}_{j,0}$. *The inverse* $\boldsymbol{G}_j$ *of* $\boldsymbol{M}_j = (\boldsymbol{M}_{j,0}, \boldsymbol{M}_{j,1})$ *is given by*

$$\boldsymbol{G}_j = \begin{bmatrix} \tilde{\boldsymbol{M}}_{j,0}^T \\ \check{\boldsymbol{G}}_{j,1} \end{bmatrix}$$

   where $\check{\boldsymbol{G}}_{j,1}$ *is defined by* (1.131) *and* $\boldsymbol{M}_j$ *and* $\boldsymbol{G}_j$ *are uniformly banded.*
2. *The scaling systems*

$$\Xi_j := \{\xi_{j,k} : k \in I_j\}, \qquad \tilde{\Xi}_j := \{\tilde{\xi}_{j,k} : k \in I_j\}, \tag{1.133}$$

   are exact, i.e., $\mathcal{P}_{d-1} \subset \operatorname{span} \Xi_j$, $\mathcal{P}_{\tilde{d}-1} \subset \operatorname{span} \tilde{\Xi}_j$, *for all* $j \geq j_0 + 1$.
3. *Setting*

$$\Upsilon_j^T := \Xi_{j+1}^T \boldsymbol{M}_{j,1}, \quad \tilde{\Upsilon}_j^T = \tilde{\Xi}_{j+1}^T \check{\boldsymbol{G}}_{j,1}^T \tag{1.134}$$

   and

$$\Upsilon := \Xi_{j_0} \cup \bigcup_{j=j_0}^{\infty} \Upsilon_j, \quad \tilde{\Upsilon} := \tilde{\Xi}_{j_0} \cup \bigcup_{j=j_0}^{\infty} \tilde{\Upsilon}_j \tag{1.135}$$

   then $\Upsilon = \{\eta_{j,k} : k \in J_j\}$, $\tilde{\Upsilon} = \{\tilde{\eta}_{j,k} : k \in J_j\}$ *are biorthogonal Riesz bases for* $L_2(0,1)$. *In particular, we have for* $\Upsilon_{j_0-1} := \Xi_{j_0}$, $\tilde{\Upsilon}_{j_0-1} := \tilde{\Xi}_{j_0}$

$$(\Upsilon_j, \tilde{\Upsilon}_{j'})_{0,(0,1)} = \delta_{j,j'} \boldsymbol{I}^{(2^j)}, \quad j, j' \geq j_0 - 1, \tag{1.136}$$

   and

$$\operatorname{diam}(\operatorname{supp} \eta_{j,k}), \operatorname{diam}(\operatorname{supp} \tilde{\eta}_{j,k}) \sim 2^{-j}, \quad j \geq j_0. \tag{1.137}$$

4. Let $\tilde{\gamma} := \sup\{s > 0 : \tilde{\xi} = {}_{d,\tilde{d}}\tilde{\xi} \in H^s(\mathbb{R})\}$ *then for* $s \in (\tilde{\gamma}, \gamma)$, *we have*

$$\Big(\sum_{k \in I_{j_0}} \big|(v, \tilde{\Xi}_{j_0,k})_{0,(0,1)}\big|^2 + \sum_{j=j_0}^{\infty} \sum_{k=1}^{2^j} 2^{2sj} \big|(v, \tilde{\eta}_{j,k})_{0,(0,1)}\big|^2 \Big)^{1/2} \sim \|v\|_{s,(0,1)} \,.$$

5. The wavelets have vanishing moments, i.e.,

$$\int_0^1 x^m \, \eta_{j,k}(x)\, dx = 0, \quad 0 \le m \le \tilde{d} - 1,$$
$$\int_0^1 x^m \, \tilde{\eta}_{j,k}(x)\, dx = 0, \quad 0 \le m \le d - 1,$$

for all $j \ge j_0 + 1$ *and* $k \in J_j$. □

We illustrate the wavelets arising by the above construction for our example $d = 3$, $\tilde{d} = 5$ in Fig. 1.10.

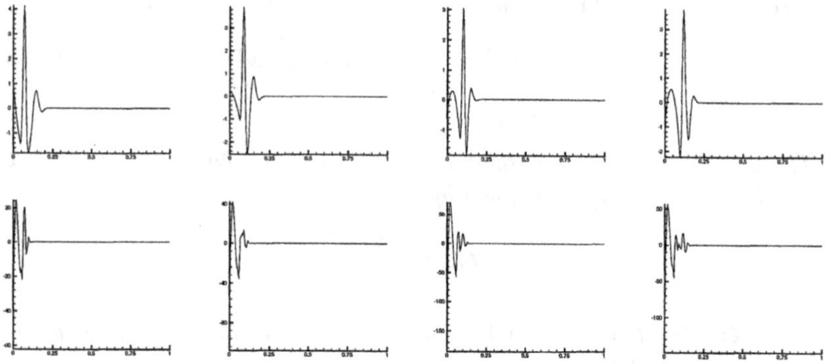

Fig. 1.10. Primal (first row) and dual (second row) wavelets for $d = 3$, $\tilde{d} = 5$ at the left boundary.

Remark 3. In some applications it might be important to start the multilevel decomposition with a very coarse initial level $j < j_0$. One should keep in mind that the concept of biorthogonality is primarily asymptotic in nature. For instance, it affects the validity of norm equivalences as well as moment conditions which become relevant when j gets large. Thus, for finitely many low levels one could always resort to simple decompositions of the primal spline spaces only. For example, in the case $d = 2$, hierarchical bases [119] would provide simple splittings for levels $j < j_0$. Moreover, one could also use non-dyadic subdivisions on few coarse levels.

We also refer to [14] for a construction of (orthogonal) wavelet bases on $(0, 1)$ on coarse scales. □

1.3.6 Quantitative Aspects of the Biorthogonalization

A first implementation of the above construction revealed the following problems concerning the quantitative properties of the arising bases. Recall that, whenever a collection $\Theta = \{\theta_\lambda : \lambda \in \mathcal{J}\}$ is a Riesz basis of the closure of its span, one has

$$c\|\mathbf{c}\|_{\ell_2(\mathcal{J})} \leq \left\| \sum_{\lambda\in\mathcal{J}} c_\lambda \theta_\lambda \right\|_{L_2} \leq C\|\mathbf{c}\|_{\ell_2(\mathcal{J})}, \qquad \mathbf{c} = (c_\lambda)_{\lambda\in\mathcal{J}} \in \ell_2(\mathcal{J}),$$

and the supremum of all c and the infimum of all C for which the above relation holds are called *Riesz constants* of Θ. The ratio of the upper and lower constant is called the *condition* of Θ.

We have noticed the following:

(a) Although Theorem 3 establishes that $\mathbf{\Gamma}_{j,X}$ is nonsingular, we have observed in several examples that $\mathrm{cond}(\mathbf{\Gamma}_{j,X}) \gg 1$. Consequently, also the entries of the mask matrices $\tilde{\boldsymbol{M}}_{j,0}$, $\boldsymbol{M}_{j,1}$ loose relative accuracy.
(b) Already the plots of our example in Figs. 1.7 and 1.8 on p. 24 indicate that the Riesz constants of Ξ_j and $\tilde{\Xi}_j$, although uniformly bounded, are considerably large. A similar behavior has been observed for the corresponding wavelet bases.

In order to circumvent these problems, we propose the following strategy. First we wish to improve the condition of the single-scale generator bases $\Xi_j, \tilde{\Xi}_j$. Since the monomial bases are increasingly ill-conditioned, one expects that the collections $\Xi_j^X := \{\xi_{j,k}^X : k \in I_j^X\}$, $X \in \{L, R\}$, and their dual analogs $\tilde{\Xi}_j^X$ of boundary functions from (1.68) and (1.73) on p. 25 inherit this property. This suggests starting from well-conditioned polynomial bases $\Pi = \{p_i : i = 0, \ldots, d-1\}$, $\tilde{\Pi} = \{\tilde{p}_i : i = 0, \ldots, \tilde{d}-1\}$ of $\mathcal{P}_d(0,1)$ and $\mathcal{P}_{\tilde{d}}(0,1)$, respectively, given by

$$p_i(x) = \sum_{r=0}^{d-1} z_{i,r}\, x^r, \qquad \tilde{p}_i(x) = \sum_{r=0}^{\tilde{d}-1} \tilde{z}_{i,r}\, x^r, \quad x \in [0,1].$$

Defining $\mathbf{Z}_L = (z_{r,i})_{r,i=0,\ldots,\tilde{d}-1}$, $\mathbf{Z}_R = \mathbf{Z}_L^{\updownarrow}$, and correspondingly $\tilde{\mathbf{Z}}_L = (\tilde{z}_{r,i})_{r,i=0,\ldots,\tilde{d}-1}$, $\tilde{\mathbf{Z}}_R = \tilde{\mathbf{Z}}_L^{\updownarrow}$, we make the following ansatz for the new collections of boundary generators

$$\Xi_j^{X,\mathrm{new}} = \mathbf{Z}_X\, \Xi_j^X, \qquad \tilde{\Xi}_j^{X,\mathrm{new}} = \tilde{\mathbf{Z}}_X\, \tilde{\Xi}_j^X, \quad X \in \{L, R\}. \tag{1.138}$$

The new generators have yet to be biorthogonalized. It is easy to see that the matrix $\mathbf{\Gamma}_{j,X}$ from (1.80) has to be replaced by $\mathbf{Z}_X \mathbf{\Gamma}_{j,X} \tilde{\mathbf{Z}}_X^T$. Hence, the matrix $\tilde{\mathbf{C}}_X$ which defines the biorthogonalized dual generators in (1.99) is now determined by the relation

$$\mathbf{Z}_X \mathbf{\Gamma}_{j,X} \tilde{\mathbf{Z}}_X^T \tilde{\mathbf{C}}_X^T = \boldsymbol{I}, \qquad X \in \{L, R\}.$$

For good choices of Π, $\tilde{\Pi}$ one expects this to alleviate problems (a) and (b). Two possibilities suggest themselves:

1. Orthonormal polynomials have optimal L_2 condition numbers.
2. Bernstein polynomials defined by

$$B_r^d(x) := b^{-d+1}\binom{d-1}{r} x^r (b-x)^{d-1-r}, \qquad r = 0, \ldots, d-1, b > 0,$$

   are known from Computer Aided Geometric Design to be well-conditioned on $[0, b]$ relative to the supremum norm, [74].

Note that the biorthogonalization as formulated above affects only the dual generators. Thus, employing Bernstein generators defined by using Z_X defined by

$$(\mathbf{Z}_L)_{l,r} = \begin{cases} (-1)^{r-l}\binom{d-1}{r}\binom{r}{l} b^{-r}, & r \geq l, \\ 0 & \text{otherwise}, \end{cases}$$

for $l, r = 0, \ldots, d-1$, preserves homogeneous boundary conditions of all but one boundary generator on the primal side which is an advantage with regard to incorporating boundary conditions, see Sect. 1.3.7 below.

Alternatively, one can combine the change of basis directly with biorthogonalization. Consider for $\check{\Xi}_j$, $\check{\tilde{\Xi}}_j$ given in (1.68), (1.73) the new bases

$$\Xi_j^{\text{new}} = \mathbf{C}_j \check{\Xi}_j, \qquad \tilde{\Xi}_j^{\text{new}} = \tilde{\mathbf{C}}_j \check{\tilde{\Xi}}_j, \tag{1.139}$$

where, on account of the remarks preceding Theorem 3 on p. 26, the matrices $\boldsymbol{C}_j, \tilde{\boldsymbol{C}}_j$ must have the form

$$\boldsymbol{C}_j = \begin{pmatrix} \mathbf{C}_L & & \mathbf{0} \\ & \boldsymbol{I}_{2^j - 2\tilde{\ell}+1-\mu(d)} & \\ \mathbf{0} & & \mathbf{C}_L^{\updownarrow} \end{pmatrix}, \quad \tilde{\boldsymbol{C}}_j = \begin{pmatrix} \tilde{\mathbf{C}}_L & & \mathbf{0} \\ & \boldsymbol{I}_{2^j - 2\tilde{\ell}+1-\mu(d)} & \\ \mathbf{0} & & \tilde{\mathbf{C}}_L^{\updownarrow} \end{pmatrix}. \tag{1.140}$$

Thus $\mathbf{C}_L, \tilde{\mathbf{C}}_L$ have to be chosen such that $\Xi_j^{\text{new}}, \tilde{\Xi}_j^{\text{new}}$ are biorthogonal,

$$(\Xi_j^{\text{new}}, \tilde{\Xi}_j^{\text{new}})_{0,(0,1)} = \boldsymbol{I} \tag{1.141}$$

which is equivalent to

$$\boldsymbol{C}_L \mathbf{\Gamma}_L \tilde{\boldsymbol{C}}_L^T = \boldsymbol{I}. \tag{1.142}$$

Again, one could base the choice of $\mathbf{C}_L, \tilde{\mathbf{C}}_L$ on transformations $\mathbf{Z}_L$, $\tilde{\mathbf{Z}}_L$ from (1.138). Instead, we mention here the following simple version which mainly addresses problem (a). Let

$$\mathbf{\Gamma}_L = \mathbf{U}\mathbf{\Sigma}\mathbf{V}^T$$

be the singular value decomposition of $\mathbf{\Gamma}_L$ given in (1.80), that is, $\mathbf{\Sigma}$ is a diagonal and $\mathbf{U}, \mathbf{V}$ are orthogonal matrices. Defining

$$\begin{aligned} \boldsymbol{C}_L &= \mathbf{\Sigma}^{-1/2}\mathbf{U}^T \in \mathbb{R}^{\tilde{d}\times\tilde{d}}, & \boldsymbol{C}_R &= \boldsymbol{C}_L^{\updownarrow}, \\ \tilde{\boldsymbol{C}}_L &= \mathbf{\Sigma}^{-1/2}\mathbf{V}^T \in \mathbb{R}^{\tilde{d}\times\tilde{d}}, & \tilde{\boldsymbol{C}}_R &= \tilde{\boldsymbol{C}}_L^{\updownarrow}, \end{aligned} \tag{1.143}$$

implies (1.142), which, in turn, confirms (1.141). The mask matrices of the primal and dual generators (1.139) and corresponding wavelets for the transformations (1.140) are then constructed as follows:

1. compute $\boldsymbol{M}_{j,0}$ from (1.91) (p. 31) and the initial stable completion $\check{\boldsymbol{M}}_{j,1}$ and inverse $\check{\boldsymbol{G}}_j$ as in (1.130), (1.131);
2. apply the basis transformation $\mathbf{C}_j$ from (1.140) to $\check{\boldsymbol{M}}_j = (\boldsymbol{M}_{j,0}, \check{\boldsymbol{M}}_{j,1})$ and $\check{\boldsymbol{G}}_j$;
3. compute the matrix $\tilde{\boldsymbol{M}}_{j,0} := \tilde{\mathbf{C}}_{j+1}^{-T}\check{\tilde{\boldsymbol{M}}}_{j,0}\tilde{\mathbf{C}}_j^T$ with $\tilde{\mathbf{C}}_j$ given in (1.140) and $\check{\tilde{\boldsymbol{M}}}_{j,0}$ from (1.94);
4. apply Theorem 4 on p. 39 to obtain the final biorthogonal system.

Recall that this situation corresponds to $\boldsymbol{C}_j = \mathbf{I}$ and $\tilde{\boldsymbol{C}}_j$ from (1.99) on p. 32. Obviously, there are many more possibilities of constructing $\boldsymbol{C}_L, \tilde{\boldsymbol{C}}_L$ satisfying (1.142) that take additional issues into account, such as preservation of boundary conditions or Riesz constants of the wavelets. In [7], a method is introduced in order to minimize the condition number of the stiffness matrix of elliptic partial differential equations using scaling functions and wavelets on $(0, 1)$.

A detailed discussion of all these issues would go beyond the scope of this monograph. We refer the reader to [53]. Both of the options indicated above, namely the use of Bernstein basis polynomials and the singular value decomposition have been thoroughly explored in [53]. The results may be roughly summarized as follows.

1. The Riesz constants for the complement bases Υ_j are *independent* of the choice of the single-scale generator bases. We again refer to [7], where the degrees of freedom in the stable completion (namely the matrices $\boldsymbol{L}_j$ and $\boldsymbol{K}_j$) were used to optimize this condition number.
2. In almost all cases the Bernstein basis polynomials give rise to the best improvement of the condition of $\Xi_j, \tilde{\Xi}_j$. This improvement grows significantly with increasing degree of polynomial exactness.

We also refer to [82] where wavelets on $(0, 1)$ with optimal locality have been constructed. Numerical experiments have shown that the condition of these bases are comparable to those achieved in [7]. Finally, we mention [54], where also the conditioning of the multiscale transformations have been investigated.

In Fig. 1.7, we have illustrated the primal scaling functions for our example $d = 3$, $\tilde{d} = 5$ using Bernstein polynomials for the boundary adaptation. The corresponding duals are shown in Fig. 1.12.

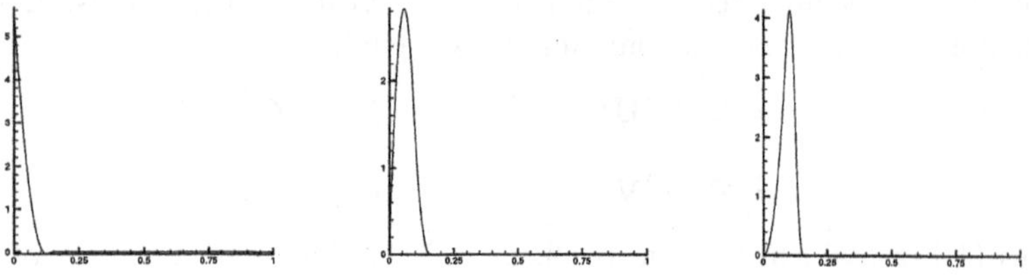

Fig. 1.11. Primal single scale functions $\xi_{5,k}$ induced by Bernstein polynomials without biorthogonalization.

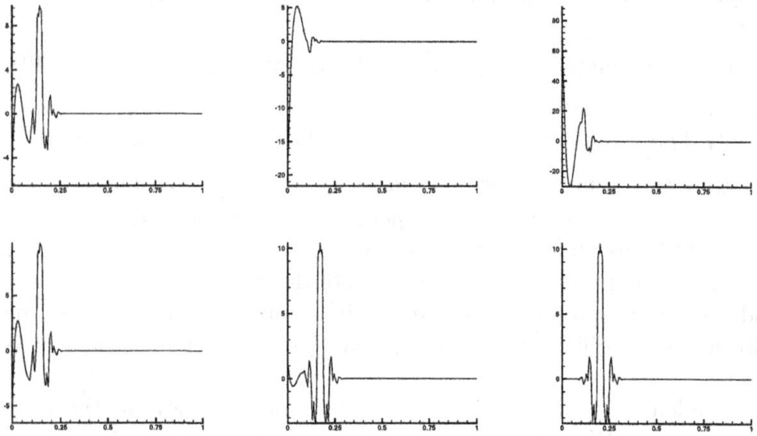

Fig. 1.12. Dual single scale functions $\tilde{\xi}_{5,k}$ induced by Bernstein polynomials.

1.3.7 Boundary Conditions

In order to use a wavelet basis to discretize a boundary value problem, the corresponding boundary conditions have to be satisfied. This can either be done indirectly e.g. by appending them in terms of Lagrange multipliers or (if possible) by incorporating them into the trial spaces.

To our knowledge, boundary conditions have been realized so far as follows:

1. no boundary conditions, i.e., the bases constructed as in the preceding sections are used;
2. homogeneous Dirichlet boundary conditions under different frameworks:
   a) homogeneous Dirichlet boundary conditions for orthonormal wavelets on $(0, 1)$ from [31] have been realized in [26];

b) homogeneous Dirichlet boundary conditions for primal *and* dual biorthogonal wavelets.
A closer look at the construction of the boundary scaling functions as in (1.66), (1.67) on p. 23 shows that all but one function in Ξ_j vanish at $x = 0$, $x = 1$, respectively. The same holds also for $\check{\Xi}_j$ in (1.73) on p. 25. One can perform the biorthogonalization in such a way that this property is preserved, see e.g. [54, 58]. Moreover, the wavelet construction can be arranged in such a way that also Υ_j and $\tilde{\Upsilon}_j$ share this property. Finally, denoting by $\xi_{j,1}^X$, $\eta_{j,1}^X$ the unique scaling function and wavelet, respectively, that does not vanish at $x = 0$, $X = L$ or $x = 1$, $X = R$, one has

$$\xi_{j,1}^X(x) = \eta_{j,1}^X(x), \qquad x = \begin{cases} 0, & \text{if } X = L, \\ 1, & \text{if } X = R. \end{cases}$$

In order to enforce homogeneous Dirichlet boundary conditions for Ξ_j, $\tilde{\Xi}_j$, we simply exclude $\xi_{j,1}^L$, $\xi_{j,1}^R$ from Ξ_j and $\tilde{\xi}_{j,1}^L$, $\tilde{\xi}_{j,1}^R$ from $\tilde{\Xi}_j$. Also only one wavelet per end point has to be modified by

$$\eta_{j,0}^X := \frac{1}{\sqrt{2}}(\eta_{j,1}^X - \xi_{j,1}^X), \quad \tilde{\eta}_{j,0}^X := \frac{1}{\sqrt{2}}(\tilde{\eta}_{j,1}^X - \tilde{\xi}_{j,1}^X), \qquad X \in \{L, R\},$$

see also Sect. 1.5 on p. 48 below;

c) complementary boundary conditions [58]: Here the primal system has homogeneous Dirichlet boundary conditions, e.g., they are in H_0^1, whereas the dual functions are not restricted. Also higher order boundary conditions can be realized.

1.3.8 Other Bases

The above construction of biorthogonal wavelets on $(0, 1)$ essentially relies on properties of B-splines. In the proof of Theorem 3 (on p. 26, see [54]) the fact was used that the space of restricted B-splines coincides with the space generated by B-splines with multiple knots at the boundaries. Moreover, it has been used that B-splines of different order are linked by certain differentiation and integration relations. In the wavelet construction, Remark 2 on p. 36 is crucial which holds for B-splines but not in general.

It is still an open problem to show all relevant properties for a construction of such general biorthogonal wavelet systems on $(0, 1)$ based on a general biorthogonal system on $\mathbb{R}$. On the other hand, there is a lack of examples of such a general biorthogonal system. Orthonormal wavelets on $(0, 1)$ have already been constructed in [31]. Prewavelets and interpolatory wavelets on $(0, 1)$ have to be constructed with the aid of other tools since they do not fit in the presented general framework. Multiwavelets on the interval can e.g. been found in [49, 85, 94]. We are not aware of other examples of biorthogonal wavelet bases on $\mathbb{R}$.

Finally, we would like to mention the modification of our above construction in [7] improving the condition number of the basis which is especially important in the performance of these bases in a Wavelet-Galerkin method. The construction of biorthogonal B-spline wavelets on $(0,1)$ in [82] differs somewhat from the presented method. In [82], boundary functions with optimal locality have been constructed. Recent numerical tests have shown that the bases in [82] behave almost as good as the modified one from [7] for a two point boundary value problem in the sense that the condition numbers of the stiffness matrices are quite close. On the other hand, the modifications in [7] increase the computational cost especially in higher dimensions.

1.4 Tensor Product Wavelets

Next, we build tensor products of the univariate systems in the following way: for any index vectors $e = (e_1, \dots, e_n)^T \in E^* := \{0,1\}^n \setminus \{\mathbf{0}\}$ and $k = (k_1, \dots, k_n)^T$, we define

$$\psi_{j,e,k}(x) := \prod_{\nu=1}^{n} \vartheta_{j,e_\nu,k_\nu}(x_\nu), \qquad \vartheta_{j,e_\nu,k_\nu} := \begin{cases} \eta_{j,k_\nu}, & e_\nu = 1, \\ \xi_{j,k_\nu}, & e_\nu = 0. \end{cases} \tag{1.144}$$

We introduce the abbreviation

$$J_{j,e} := \begin{cases} J_j, & e = 1, \\ I_j, & e = 0, \end{cases} \tag{1.145}$$

i.e., the parameter $e \in \{0,1\}$ allows to distinguish between scaling functions and wavelets. Then, k in (1.144) ranges over the set of indices

$$\mathcal{J}_j := \bigcup_{e \in E^*} \mathcal{J}_{j,e},$$

where $\mathcal{J}_{j,e} := J_{j,e_1} \times \cdots \times J_{j,e_n}$, see (1.145).

Moreover, we define the scaling functions by $\varphi_{j,k}(x) := \prod_{\nu=1}^n \xi_{j,k_\nu}(x_\nu)$, for $k \in \mathcal{I}_j := \mathcal{I}_j^n$. The wavelet systems are then defined by

$$\Psi_j := \bigcup_{e \in E^*} \Psi_{j,e}, \qquad \Psi_{j,e} := \{\psi_{j,e,k} : k \in \mathcal{J}_{j,e}\},$$

and all definitions apply similarly to the dual (biorthogonal) systems of functions.

Note that also certain non-separable, i.e., non-tensor product, wavelets for $L_2(\mathbb{R}^n)$ are known. Let us mention box-spline prewavelets [110] and bidimensional wavelets with hexagonal symmetry [34].

1.5 Wavelets on General Domains

Meanwhile wavelet bases have been constructed for a variety of operator equations such as the classical boundary integral operators on possibly closed manifolds of dimension up to 2 as well as for elliptic second order boundary value problems on domains in $\mathbb{R}^n$, [22, 23, 24, 33, 59, 60, 62]. Here we are interested in wavelet bases that are well suited in particular for the numerical treatment of second order elliptic partial differential equations, i.e., problems in $H^1(\Omega)$. As in most of the above above mentioned constructions, here, the domain of interest Ω is split into subdomains Ω_i, which are images of the reference element $\hat{\Omega} = (0,1)^n$ under appropriate parametric mappings. It is well-known that a function which is in $H^1(\Omega_i)$ for each subdomain Ω_i and which is globally continuous, belongs to $H^1(\Omega)$. Hence, we have to construct globally continuous functions. Thus, the multiresolution analysis on Ω is then easily obtained by transformations of properly matched systems on $\hat{\Omega}$. The construction of the corresponding wavelet bases is not as straightforward. In [59], the above mentioned technique of stable completions has been used to construct such wavelets, there called *composite wavelet basis*. Here, we are going to outline a different approach called *Wavelet Element Method*, [22, 23, 24]. In this approach explicit matching conditions and coefficients for the wavelets are derived in order to ensure continuity across the interelement boundaries. A similar path has also been used in [33]. We will show that the resulting bases yield norm equivalences for $H^s(\Omega)$, $-1/2 < s < 3/2$, which already covers a wide range of applications.

If one is interested in a particular problem which is not governed by this range, one has to resort to other bases constructions. Such applications could either be higher order problems, i.e., $s \geq 3/2$, e.g. enforcing globally C^1-functions, or problems on spaces with negative order, $s \leq -1/2$, such as integral equations. We would like to mention the constructions in [60, 62]. In [62], wavelet bases for Lagrange type finite elements as multiresolution have been constructed yielding to the range $|s| < 3/2$ ($|s| \leq 1$ for Lipschitz manifolds). In [60], a general theorem was used that characterizes function spaces over a domain or manifold in terms of product spaces where each factor is a corresponding local function space subject to certain boundary conditions. This approach gives rise to wavelet bases that characterize the full range of s which is permitted by the univariate functions. In particular, arbitrary smoothness and characterization for values $s \leq -1/2$ are realized.

The most general approach to the Wavelet Element Method is presented in [23], where an abstract theorem ensures the construction of such bases in any spatial dimension and for all possible geometric properties of the domain decomposition. Here we prefer to follow [24] and detail the construction for 1D, 2D and 3D separately, since this also shows how to realize the construction numerically.

We will first describe our technical assumptions on Ω and the parametric mappings from the reference domain to the subdomains. In Subsect. 1.5.2

on p. 53, we detail the construction of scaling functions and wavelets on a single subdomain. The matching for the scaling functions will be described in Subsect. 1.5.3 on p. 54. Most of the wavelets in Ω simply arise by a mapping from the reference element without any matching; these functions will be described in Subsect. 1.5.4 on p. 55. The more complex matching for the wavelets will be detailed in the subsequent sections for the 1D, 2D and 3D cases, separately. Finally, in Subsect. 1.5.8 on p. 78, certain characterization properties of Sobolev spaces are summarized.

Labeling by Grid Points. The starting point of the constructions is formed by two families of scaling functions

$$\Xi_j := \{\xi_{j,k} : k \in I_j\}, \ \tilde{\Xi}_j := \{\tilde{\xi}_{j,k} : k \in I_j\} \subset L_2(0,1),$$

where again I_j denotes an appropriate set of indices. For subsequent convenience, these functions will not be labeled by integers as before, but rather by a set of real indices

$$I_j := \{\tau_{j,1}, \ldots, \tau_{j,K_j}\}, \qquad 0 = \tau_{j,1} < \tau_{j,2} < \cdots < \tau_{j,K_j} = 1. \tag{1.146}$$

In other words, each basis function is associated with a *node*, or grid point, in the interval $[0,1]$; the actual position of the internal nodes $\tau_{j,2}, \ldots, \tau_{j,K_j-1}$ will be irrelevant in the sequel, except that it is required that $I_j \subset I_{j+1}$.

We assume the following properties of the scaling function systems:

Assumption 1. *a.) The systems Ξ_j and $\tilde{\Xi}_j$ are* refinable

b.) The functions have local support: diam (supp $\xi_{j,k}$) $\sim$ diam (supp $\tilde{\xi}_{j,k}$) $\sim 2^{-j}$.

c.) The systems are biorthogonal, *i.e.,* $(\xi_{j,k}, \tilde{\xi}_{j,k'})_{0,(0,1)} = \delta_{k,k'}$ *for all* $k, k' \in I_j$.

c.) The systems Ξ_j, $\tilde{\Xi}_j$ are uniformly stable.

d.) The functions are regular, *i.e.,* $\xi_{j,k} \in H^{\gamma}(0,1)$, $\tilde{\xi}_{j,k} \in H^{\tilde{\gamma}}(0,1)$, *for some* $\gamma, \tilde{\gamma} > 1$.

e.) The systems are exact *of order* L, $\tilde{L} \geq 1$.

f.) There exist biorthogonal complement spaces W_j and $\tilde{W}_j$ which have bases

$$\Upsilon_j = \{\eta_{j,h} : h \in J_j\}, \qquad \tilde{\Upsilon}_j = \{\tilde{\eta}_{j,h} : h \in J_j\},$$

which are biorthogonal and uniformly stable.

The wavelet systems Υ_j and $\tilde{\Upsilon}_j$ are also labeled by grid points

$$J_j := I_{j+1} \setminus I_j = \{\nu_{j,1}, \ldots, \nu_{j,M_j}\}, \quad 0 < \nu_{j,1} < \cdots < \nu_{j,M_j} < 1.$$

We assume that systems Ξ_j and Υ_j are *boundary adapted*, i.e., at each boundary point:

(i) only one basis function in each system is not vanishing; precisely,

$$\xi_{j,k}(0) \neq 0 \iff k = 0, \qquad \xi_{j,k}(1) \neq 0 \iff k = 1, \tag{1.147}$$
$$\eta_{j,h}(0) \neq 0 \iff h = \nu_{j,1}, \qquad \eta_{j,h}(1) \neq 0 \iff h = \nu_{j,M_j} \tag{1.148}$$

(ii) the non-vanishing scaling and wavelet functions take the same value; precisely, there exist constants c_0 and c_1 independent of j such that

$$\xi_{j,0}(0) = \eta_{j,\nu_{j,1}}(0) = c_0 2^{j/2}, \qquad \xi_{j,1}(1) = \eta_{j,\nu_{j,M_j}}(1) = c_1 2^{j/2}. \tag{1.149}$$

The same holds for the dual systems $\tilde{\Xi}_j$ and $\tilde{\Upsilon}_j$. Moreover, the system Ξ_j is required to be *boundary symmetric*, i.e.,

$$\xi_{j,0}(0) = \xi_{j,1}(1) =: \lambda_j \tag{1.150}$$

and also for $\tilde{\Xi}_j$. The following concept will be important in the sequel. The system Ξ_j is said to be *reflection invariant*, if I_j is invariant under the mapping $x \mapsto 1 - x$ and

$$\xi_{j,k}(1-x) = \xi_{j,1-k}(x), \qquad \text{for all } x \in [0,1] \text{ and } k \in I_j. \tag{1.151}$$

An analogous definition can be given for the system Υ_j, as well as for the dual systems. If Ξ_j is reflection invariant, then Υ_j can be built to have the same property. This will be always implicitly assumed.

Boundary adapted generator and wavelet systems can be easily modified to fulfill homogeneous Dirichlet boundary conditions. To this end, let us first introduce the following sets of the *internal* grid points:

$$I_j^{int} := I_j \setminus \{0,1\}, \qquad J_j^{int} := J_j \setminus \{\nu_{j,1}, \nu_{j,M_j}\}. \tag{1.152}$$

Let us collect in the vector $\beta = (\beta_0, \beta_1) \in \{0,1\}^2$ the information about where homogeneous boundary conditions are enforced, i.e., $\beta_x = 1$ means no boundary condition, whereas $\beta_x = 0$ denotes boundary condition at the point $x \in \{0,1\}$. Correspondingly, let us set

$$I_j^\beta := \begin{cases} I_j^{int}, & \text{if } \beta = (0,0), \\ I_j \setminus \{0\}, & \text{if } \beta = (0,1), \\ I_j \setminus \{1\}, & \text{if } \beta = (1,0), \\ I_j, & \text{if } \beta = (1,1). \end{cases} \tag{1.153}$$

Let the generator systems be defined as

$$\Xi_j^\beta := \{\xi_{j,k} : k \in I_j^\beta\}, \qquad \tilde{\Xi}_j^\beta := \{\tilde{\xi}_{j,k} : k \in I_j^\beta\},$$

and let us define the multiresolution analyses

$$V_j^\beta := \operatorname{span} \Xi_j^\beta, \qquad \tilde{V}_j^\beta := \operatorname{span} \tilde{\Xi}_j^\beta. \tag{1.154}$$

Note that we have simply omitted the scaling functions which do not vanish at those end points of the interval where boundary conditions are enforced.

The associated biorthogonal wavelet systems Υ_j^β, $\tilde{\Upsilon}_j^\beta$ are the same as the previously defined ones, except that we possibly change the first and/or the last wavelet depending on β. More precisely, the wavelets can be chosen to vanish at each boundary point in which the corresponding component of β is zero. If the boundary condition is prescribed at 0, the first wavelets $\eta_{j,\nu_{j,1}}$ and $\tilde{\eta}_{j,\nu_{j,1}}$ are replaced by

$$\eta^D_{j,\nu_{j,1}} := \frac{1}{\sqrt{2}}(\eta_{j,\nu_{j,1}} - \xi_{j,0}), \qquad \tilde{\eta}^D_{j,\nu_{j,1}} := \frac{1}{\sqrt{2}}(\tilde{\eta}_{j,\nu_{j,1}} - \tilde{\xi}_{j,0}), \tag{1.155}$$

respectively. The wavelets $\eta^D_{j,\nu_{j,M_j}}$ and $\tilde{\eta}^D_{j,\nu_{j,M_j}}$ vanishing at 1 are defined similarly. Observe that the set of grid points J_j^β which labels the wavelets does not change, i.e., $J_j^\beta = J_j$ for all choices of β.

The new systems Ξ_j^β, Υ_j^β and $\tilde{\Xi}_j^\beta$, $\tilde{\Upsilon}_j^\beta$ fulfill the conditions of a multiresolution analysis. In addition, if the systems Ξ_j and $\tilde{\Xi}_j$ are reflection invariant, see (1.151), the systems with boundary conditions can be built to be reflection invariant as well, in an obvious sense (i.e., the mapping $x \mapsto 1-x$ induces a mapping of Ξ_j^β into itself if $\beta = (0,0)$ or $\beta = (1,1)$, while it produces an exchange of $\Xi_j^{(0,1)}$ with $\Xi_j^{(1,0)}$ in the other cases).

Tensor products. Next, we build multivariate generators and wavelets from univariate ones using tensor products. Hereafter, we describe the construction in the domain $\hat{\Omega} = (0,1)^n$, which will serve as a reference domain later on.

Let the vector $b = (\beta^1, \ldots, \beta^n)$ (where $\beta^l \in \{0,1\}^2$ for $1 \le l \le n$) contain the information on the boundary conditions to be enforced. Let us set, for all $j \ge j_0$,

$$S_j^b(\hat{\Omega}) := V_j^{\beta^1} \otimes \cdots \otimes V_j^{\beta^n}, \tag{1.156}$$

and similarly for $\tilde{S}_j^b(\hat{\Omega})$. These spaces are trivially nested. In order to construct a basis for them, we define for $\hat{x} = (\hat{x}_1, \ldots, \hat{x}_n) \in \hat{\Omega}$ and $\hat{k} = (\hat{k}_1, \ldots, \hat{k}_n) \in \mathcal{I}_j^b := I_j^{\beta^1} \times \cdots \times I_j^{\beta^n}$

$$\hat{\varphi}_{j,\hat{k}}(\hat{x}) := (\xi_{j,\hat{k}_1} \otimes \cdots \otimes \xi_{j,\hat{k}_n})(\hat{x}) = \prod_{l=1}^n \xi_{j,\hat{k}_l}(\hat{x}_l);$$

we set

$$\hat{\Phi}_j := \{\hat{\varphi}_{j,\hat{k}} : \hat{k} \in \mathcal{I}_j^b\},$$

so that

$$S_j^b(\hat{\Omega}) = \operatorname{span}(\hat{\Phi}_j), \qquad \tilde{S}_j^b(\hat{\Omega}) = \operatorname{span}(\hat{\tilde{\Phi}}_j).$$

Biorthogonal complement spaces $W_j^b(\hat{\Omega})$, i.e., spaces satisfying

$$S^b_{j+1}(\hat{\Omega}) = S^b_j(\hat{\Omega}) \oplus W^b_j(\hat{\Omega}), \qquad W^b_j(\hat{\Omega}) \perp \tilde{S}^b_j(\hat{\Omega}),$$

are defined as follows. Set $\mathcal{J}^b_j := \mathcal{I}^b_{j+1} \setminus \mathcal{I}^b_j$ and for any $\hat{h} = (\hat{h}_1, \ldots, \hat{h}_n) \in \mathcal{J}^b_j$, define the corresponding wavelet

$$\hat{\psi}_{j,\hat{h}}(\hat{x}) := (\hat{\vartheta}_{\hat{h}_1} \otimes \cdots \otimes \hat{\vartheta}_{\hat{h}_n})(\hat{x}) \;=\; \prod_{l=1}^{n} \hat{\vartheta}_{\hat{h}_l}(\hat{x}_l),$$

where

$$\hat{\vartheta}_{\hat{h}_l} := \begin{cases} \xi_{j,\hat{h}_l}, & \text{if } \hat{h}_l \in I^{\beta^l}_j, \\ \eta_{j,\hat{h}_l}, & \text{if } \hat{h}_l \in J_j. \end{cases}$$

Then, set

$$\hat{\Psi}_j := \{\hat{\psi}_{j,\hat{h}} : \hat{h} \in \mathcal{J}^b_j\}, \qquad \text{and} \qquad W_j(\hat{\Omega}) := \operatorname{span} \hat{\Psi}_j.$$

The definition of $\hat{\tilde{\Psi}}_j$ and $\tilde{W}^b_j(\hat{\Omega})$ are similar. The wavelet systems $\hat{\Psi}_j$ and $\hat{\tilde{\Psi}}_j$ are biorthogonal and form Riesz bases in $L_2(\hat{\Omega})$: the norm equivalences (1.4) on p. 2 extend to the multivariate case as well.

Considering the boundary values, we note that, given any $l \in \{1, \ldots, n\}$ and $x \in \{0, 1\}$, we have

$$(\hat{\varphi}_{j,\hat{k}})_{|\hat{x}_l = x} \equiv 0 \qquad \text{iff} \qquad \hat{k}_l \neq x \;\text{ or }\; (\hat{k}_l = x \text{ and } \beta^l_x = 0)$$

and

$$(\hat{\psi}_{j,\hat{h}})_{|\hat{x}_l = x} \equiv 0 \qquad \text{iff} \qquad \hat{h}_l \neq \nu_x \;\text{ or }\; (\hat{h}_l = \nu_x \text{ and } \beta^l_x = 0),$$

with $\nu_d = \nu_{j,1}$ if $x = 0$, and $\nu_d = \nu_{j,M_j}$ if $x = 1$.

1.5.1 Domain Decomposition and Parametric Mappings

Let us consider our domain of interest $\Omega \subset \mathbb{R}^n$. The boundary $\partial\Omega$ is subdivided in two relatively open parts (with respect to $\partial\Omega$), the Dirichlet part Γ_{Dir} and the Neumann part Γ_{Neu}, in such a way that

$$\overline{\partial\Omega} = \bar{\Gamma}_{\mathrm{Dir}} \cup \bar{\Gamma}_{\mathrm{Neu}}, \qquad \Gamma_{\mathrm{Dir}} \cap \Gamma_{\mathrm{Neu}} = \emptyset. \tag{1.157}$$

We assume that there exist N open disjoint subdomains $\Omega_i \subseteq \Omega$ ($i = 1, \ldots, N$) such that

$$\bar{\Omega} = \bigcup_{i=1}^{N} \bar{\Omega}_i \tag{1.158}$$

and such that, for some $r \geq \gamma$, there exist r-time continuously differentiable mappings $F_i : \bar{\hat{\Omega}} \to \bar{\Omega}_i$ ($i = 1, \ldots, N$) satisfying

$$\Omega_i = F_i(\hat{\Omega}), \qquad |JF_i| := \det(JF_i) > 0 \text{ in } \bar{\hat{\Omega}}, \tag{1.159}$$

where JF_i denotes the Jacobian of F_i; in the sequel, it will be useful to set $G_i := F_i^{-1}$.

Let us first set some notation, starting with the reference domain $\hat{\Omega}$. For $0 \le p \le n-1$, a *p-face* of $\hat{\Omega}$ is a subset $\hat{\sigma} \subset \partial\hat{\Omega}$ defined by the choice of a set $\mathcal{L}_{\hat{\sigma}}$ of different indices $l_1, \dots, l_{n-p} \in \{1, \dots, n\}$ and a set of integers $d_1, \dots, d_{n-p} \in \{0, 1\}$ in the following way

$$\hat{\sigma} = \{(\hat{x}_1, \dots, \hat{x}_n) \ : \ \hat{x}_{l_1} = d_1, \ \dots, \ \hat{x}_{l_{n-p}} = d_{n-p}, \text{ and } 0 \le \hat{x}_l \le 1 \text{ if } l \notin \mathcal{L}_{\hat{\sigma}}\} \tag{1.160}$$

(thus, e.g., in 3D, a 0-face is a vertex, a 1-face is a side and a 2-face is a usual face of the reference cube). The coordinates $\hat{x}_l$ with $l \in \mathcal{L}_{\hat{\sigma}}$ will be termed the *frozen coordinates* of $\hat{\sigma}$, whereas the remaining coordinates will be termed the *free coordinates* of $\hat{\sigma}$.

The image of a p-face of $\hat{\Omega}$ under the mapping F_i will be termed a p-face of Ω_i; if $\Gamma_{i,i'} := \partial\Omega_i \cap \partial\Omega_{i'}$ is nonempty for some $i \neq i'$, then we assume that $\Gamma_{i,i'}$ is a p-face of both Ω_i and $\Omega_{i'}$ for some $0 \le p \le n-1$.

Finally, the decomposition is assumed to be *geometrically conforming* in the following sense: the intersection $\bar{\Omega}_i \cap \bar{\Omega}_{i'}$ for $i \neq i'$ is either empty or a p-face, $0 \le p \le n-1$; moreover, for $i = 1, \dots, N$ we suppose that $\partial\Omega_i \cap \bar{\Gamma}_{\mathrm{Dir}}$ and $\partial\Omega_i \cap \bar{\Gamma}_{\mathrm{Neu}}$ are (possibly empty) unions of p-faces of Ω_i.

Next, we set up the technical assumptions for the mappings F_i. To formulate these, we need some notation. Let $\hat{\sigma}$ and $\hat{\sigma}'$ be two p-faces of $\hat{\Omega}$, and let $H : \hat{\sigma} \to \hat{\sigma}'$ be a bijective mapping. We shall say that H is *order-preserving* if it is permutation of the free coordinates of $\hat{\sigma}$. An order-preserving mapping is a particular case of an affine mapping (see [23], Lemma 4.1).

In addition, setting

$$\Gamma_{i,i'} = F_i(\hat{\sigma}) = F_{i'}(\hat{\sigma}'), \tag{1.161}$$

with two p-faces $\hat{\sigma}$ and $\hat{\sigma}'$ of $\hat{\Omega}$, we require that the bijection

$$H_{i,i'} := G_{i'} \circ F_i \ : \hat{\sigma} \to \hat{\sigma}' \equiv \big(G_{i'} \circ (F_i|_{\hat{\sigma}})\big)|_{F_i(\hat{\sigma})} \tag{1.162}$$

fulfills the following

Assumption 2. *a.) $H_{i,i'}$ is affine;*
b.) in addition, if the systems of scaling functions and wavelets on $(0,1)$ are not *reflection invariant, then $H_{i,i'}$ is order-preserving.*

Let us add some comments on the latter assumption. One can show (see e.g. [23]) that a function $H_{i,i'}$ is affine if and only if it is a composition of elementary permutations and reflections of the free coordinates. This shows that Assumption 2 is more of technical nature since we assume that the parameterization of p-faces does not change enormously when going from one subdomain to an adjacent one. Since in practice often parameterizations of the interfaces are given by the user (and then F_i is computed thereof, see e.g. [78, 79]) a.) is no restriction. Also b.) is easy to ensure in many situations. For example, biorthogonal B-splines can be symmetrized with respect to $[0,1]$ also for odd values of d (see [58]) and hence these systems are in particular reflection invariant. Moreover, if a.) holds true, it can easily be seen that for $n=2$ it is always possible to modify the mappings F_i in such a way that for the new mapping all $H_{i,i'}$ are order-preserving. However, this is not true if $n=3$, as the example of a 3D Moebius ring indicates.

To summarize, we need the following assumptions:

(a) The mappings $F_i : \bar{\hat{\Omega}} \to \bar{\Omega}_i$ are r-times continuously differentiable, with $r \geq \gamma$, and satisfy $|JF_i| > 0$ in $\bar{\hat{\Omega}}$;
(b) The mappings F_i fulfill Assumption 2 above;
(c) The domain decomposition is geometrically conforming.

1.5.2 Multiresolution and Wavelets on the Subdomains

Let us now introduce multiresolution analyses on each Ω_i, $i=1,\ldots,N$, by "mapping" appropriate multiresolution analyses on $\hat{\Omega}$. To this end, let us define the vector $b(\Omega_i) = (\beta^1,\ldots,\beta^n) \in \{0,1\}^{2n}$ as follows

$$\beta^l_x = \begin{cases} 0, & \text{if } F_i(\{\hat{x}_l = x\}) \subset \Gamma_{\mathrm{Dir}}, \\ 1, & \text{otherwise}, \end{cases} \qquad l=1,\ldots,n, \quad x=0,1. \tag{1.163}$$

Moreover, let us introduce the one-to-one transformation $v \mapsto \hat{v} := v \circ F_i$, which maps functions defined in $\bar{\Omega}$ into functions defined in $\bar{\hat{\Omega}}$. Next, for all $j \geq j_0$, let us set

$$S_j(\Omega_i) := \{v : \hat{v} \in S_j^{b(\Omega_i)}(\hat{\Omega})\}, \tag{1.164}$$

see (1.156). A basis in this space is obtained as follows. For any $\hat{k} \in \mathcal{I}_j^{b(\Omega_i)}$, set $k^{(i)} := F_i(\hat{k})$; then, define the set of grid points

$$\mathcal{K}_j^i := \{k^{(i)} : \hat{k} \in \mathcal{I}_j^{b(\Omega_i)}\} \tag{1.165}$$

in $\bar{\Omega}_i$. The grid point k, whose image under G_i is $\hat{k}$, is associated to the function in $V_j(\Omega_i)$

$$\varphi_{j,k}^{(i)} := \hat{\varphi}_{j,\hat{k}} \circ G_i, \tag{1.166}$$

i.e., $\widehat{\varphi_{j,k}^{(i)}} = \hat{\varphi}_{j,\hat{k}}$. The set of all these functions will be denoted by Φ_j^i. This set and the dual set $\tilde{\Phi}_j^i$ form biorthogonal bases of $S_j(\Omega_i)$ and $\tilde{S}_j(\Omega_i)$, respectively, with respect to the (non-standard) inner product in $L_2(\Omega_i)$

$$\langle u,v\rangle_{\Omega_i} := \int_{\Omega_i} u(x)\, v(x)\, |JG_i(x)|\, dx = \int_{\hat{\Omega}} \hat{u}(\hat{x})\, \hat{v}(\hat{x})\, d\hat{x}, \tag{1.167}$$

which, due to the properties of the transformation of the domains, induces an equivalent L_2-type norm

$$\|v\|_{0,\Omega_i}^2 \sim \langle v, v\rangle_{\Omega_i} = \|\hat{v}\|_{0,\hat{\Omega}}^2, \qquad v \in L_2(\Omega_i). \tag{1.168}$$

Coming to the detail spaces, a complement of $S_j(\Omega_i)$ in $S_{j+1}(\Omega_i)$ can be defined as

$$W_j(\Omega_i) := \{w : \hat{w} \in W_j^{b(\Omega_i)}(\hat{\Omega})\}. \tag{1.169}$$

A basis Ψ_j^i in this space is associated to the grid

$$\mathcal{H}_j^i := \mathcal{K}_{j+1}^i \setminus \mathcal{K}_j^i = \{h = F_i(\hat{h}) : \hat{h} \in \mathcal{J}_j^{b(\Omega_i)}\} \tag{1.170}$$

through the relation

$$\psi_{j,h}^{(i)} := \hat{\psi}_{j,\hat{h}} \circ G_i, \qquad h \in \mathcal{H}_j^i, \quad h = F_i(\hat{h}). \tag{1.171}$$

The space $W_j(\Omega_i)$ and the similarly defined space $\tilde{W}_j(\Omega_i)$ form biorthogonal complements; the bases Ψ_j^i, and $\tilde{\Psi}_j^i$ are biorthogonal (with respect to $\langle\cdot,\cdot\rangle_{\Omega_i}$). It is easily seen that the dual multiresolution analyses on Ω_i defined in this way inherit the properties of the multiresolution analyses on $\hat{\Omega}$.

Finally, we introduce a concept that will be useful in the sequel. A point $h \in \mathcal{H}_j^i$ is termed *internal* to Ω_i if $h = F_i(\hat{h})$, with $\hat{h} = (\hat{h}_1, \cdots, \hat{h}_n)$ such that each component $\hat{h}_l$ belongs to $I_j^{int} \cup J_j^{int}$, (1.152) on p. 49.

1.5.3 Multiresolution on the Global Domain Ω

Now we describe the construction of dual multiresolution analyses on $\bar{\Omega}$. Let us define, for all $j \geq j_0$,

$$S_j(\Omega) := \{v \in C^0(\bar{\Omega}) : v_{|\Omega_i} \in S_j(\Omega_i), i = 1, \ldots, N\}; \tag{1.172}$$

the dual spaces $\tilde{S}_j(\Omega)$ are defined in a similar manner. In order to define a basis of $S_j(\Omega)$, let us introduce the set

$$\mathcal{K}_j := \bigcup_{i=1}^{N} \mathcal{K}_j^i \tag{1.173}$$

containing all the grid points in $\bar{\Omega}$. Each point of $\mathcal{K}_j$ can be associated to one single scale basis function of $S_j(\Omega)$, and conversely. To accomplish this, let us set

$$I(k) := \left\{i \in \{1, \ldots, N\} : k \in \bar{\Omega}_i\right\}, \quad k \in \mathcal{K}_j, \tag{1.174}$$

as well as

$$\hat{k}^{(i)} := G_i(k), \quad i \in I(k), \quad k \in \mathcal{K}_j. \tag{1.175}$$

Then, for any $k \in \mathcal{K}_j$ let us define the function $\varphi_{j,k}$ as follows

$$\varphi_{j,k\,|\Omega_i} := \begin{cases} |I(k)|^{-1/2}\,\varphi_{j,k}^{(i)}, & \text{if } i \in I(k), \\ 0, & \text{otherwise.} \end{cases} \tag{1.176}$$

This function belongs to $S_j(\Omega)$, since it is continuous across the interelement boundaries (see [23], Sect. 4.2). Let us now set $\Phi_j := \{\varphi_{j,k} : k \in \mathcal{K}_j\}$. The dual family $\tilde{\Phi}_j := \{\tilde{\varphi}_{j,k} : k \in \mathcal{K}_j\}$ is defined as in (1.176), simply by replacing each $\varphi_{j,k}^{(i)}$ by $\tilde{\varphi}_{j,k}^{(i)}$. Then, we have $S_j(\Omega) = \operatorname{span}\Phi_j$, $\tilde{S}_j(\Omega) = \operatorname{span}\tilde{\Phi}_j$. By defining the L_2-type inner product on Ω

$$\langle u, v\rangle_\Omega := \sum_{i=1}^{N} \langle u, v\rangle_{\Omega_i}, \tag{1.177}$$

it is easy to obtain the biorthogonality relations $\langle \varphi_{j,k}, \tilde{\varphi}_{j,k'}\rangle_\Omega = \delta_{k,k'}$, from those in each Ω_i.

1.5.4 Wavelets on the Global Domain

We now begin the construction of biorthogonal complement spaces $W_j(\Omega)$ and $\tilde{W}_j(\Omega)$ $(j \geq j_0)$ for $S_{j+1}(\Omega)$ and $\tilde{S}_{j+1}(\Omega)$ as well as the corresponding biorthogonal bases Ψ_j and $\tilde{\Psi}_j$. Given the set of grid points

$$\mathcal{H}_j := \mathcal{K}_{j+1} \setminus \mathcal{K}_j =: \bigcup_{i=1}^{N} \mathcal{H}_j^i, \tag{1.178}$$

we shall associate to each $h \in \mathcal{H}_j$ a function $\psi_{j,h} \in W_j(\Omega)$ and a function $\tilde{\psi}_{j,h} \in \tilde{W}_j(\Omega)$. Then, we shall set $\Psi_j := \{\psi_{j,h} : h \in \mathcal{H}_j\}$, $\tilde{\Psi}_j := \{\tilde{\psi}_{j,h} : h \in \mathcal{H}_j\}$.

At first, let us observe that if $h \in \mathcal{H}_j^i$ is such that the associated local wavelet $\psi_{j,h}^{(i)}$ (defined in (1.171)) vanishes identically on $\partial\Omega_i \setminus \partial\Omega$, then the function

$$\psi_{j,h}(x) := \begin{cases} \psi_{j,h}^{(i)}(x), & \text{if } x \in \Omega_i, \\ 0, & \text{otherwise,} \end{cases} \tag{1.179}$$

will be a wavelet on Ω associated to h. This situation occurs either when h is an internal point of Ω_i (recall the definition of internal point given at the end of the previous subsection), or when all non internal coordinates of h correspond to a physical boundary (see Fig. 1.13).

The remaining wavelets will be obtained by matching suitable combinations of scaling functions and wavelets in contiguous domains. In [23], a general procedure was introduced that works for each spatial dimension n. It was proven there that this method gives rise to a wavelet basis with the desired properties. Since this general case involves some technicalities, we restrict

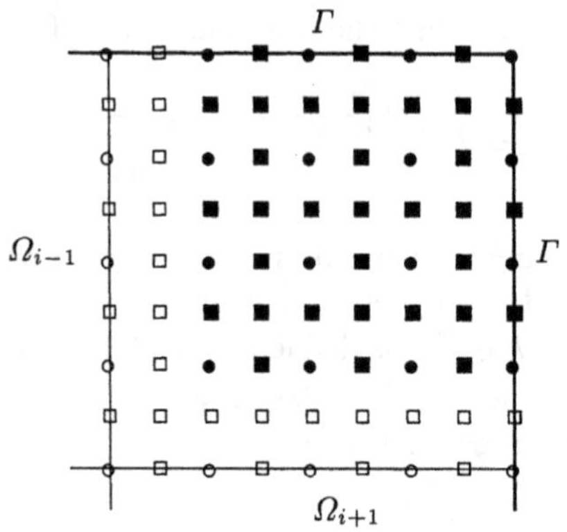

Fig. 1.13. Scaling function grid points (circles) and wavelet grid points (boxes) in the subdomain Ω_i (here for simplicity rectangular). The upper and right parts of the boundary of the subdomain belong to Γ. The filled wavelet grid points are associated to wavelets on Ω constructed according to (1.179).

ourselves here to the relevant cases $n = 1, 2, 3$. In each of the three coming subsections, we shall detail the construction of the univariate, bivariate and trivariate matched wavelets, respectively. Precisely,

- in 1D, wavelets are matched across the interface between two contiguous subdomains (i.e., subintervals);
- in 2D, wavelets are matched around the common vertex of several subdomains (cross point), or across the common side between two subdomains;
- in 3D, wavelets are matched around cross points, or around a common edge of several subdomains, or across the common face of two subdomains.

Example 1. Throughout the next subsections, we shall illustrate our construction of matched scaling functions and wavelets starting from biorthogonal spline wavelets on the real line, as introduced in [30]. The corresponding multiresolutions on the interval are built as in Sec. 1.3 (see also [53, 54]) with the choice of parameters $L = 2$ and $\tilde{L} = 4$, using SVD for the biorthogonalization. The particular implementation used to produce the pictures of the present paper is described in [9]. Figs. 1.14 and 1.15 show the primal and dual scaling functions which are boundary adapted, whereas Figs. 1.16 and 1.17 refer to primal and dual wavelets. These, and all the subsequent figures of this section, correspond to the level $j = 4$. Fig. 1.18 shows the modified wavelets defined in (1.155) for the B-spline multiresolution chosen to illustrate our construction. □

1.5.5 Univariate Matched Wavelets and Other Functions

In this section, we describe the construction of matched wavelets and other functions in the one-dimensional case. Since this material will be used in the

Fig. 1.14. Primal scaling functions.

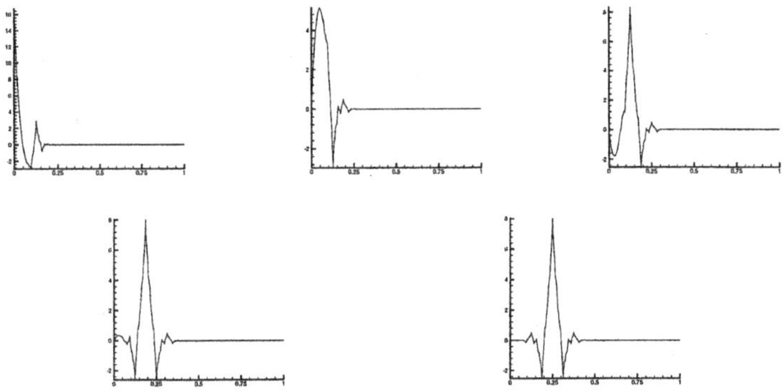

Fig. 1.15. Dual scaling functions.

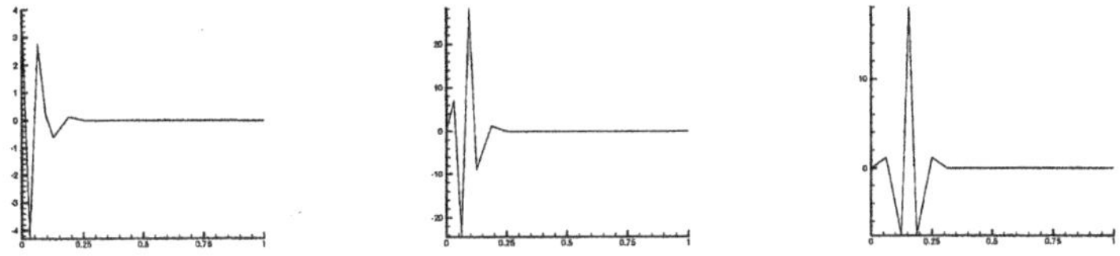

Fig. 1.16. Primal wavelets.

subsequent construction of higher dimensional wavelets, we restrict ourselves to the natural reference situation of the interval $I = (-1, 1)$ divided in the two subintervals $I_- = (-1, 0)$ and $I_+ = (0, 1)$ by the interface point $C = 0$. It is straightforward to reduce any other one-dimensional matching to the present situation, by possibly introducing a suitable parametric mapping.

The scaling function $\hat{\varphi}_{j,0}$, associated to the interface point $C = 0$, is defined by

$$\hat{\varphi}_{j,0}(\hat{x}) := \frac{1}{\sqrt{2}} \begin{cases} \xi_{j,1}(\hat{x}+1), & \hat{x} \in I_-, \\ \xi_{j,0}(\hat{x}), & \hat{x} \in I_+. \end{cases} \tag{1.180}$$

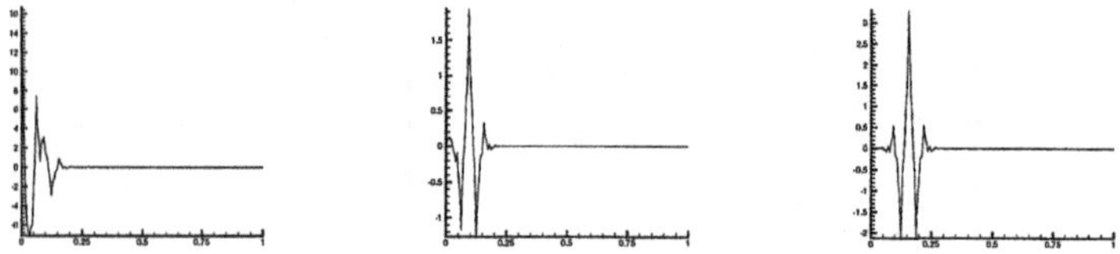

Fig. 1.17. Dual wavelets.

Fig. 1.18. Primal and dual wavelet corresponding to the first wavelet grid point having homogeneous boundary conditions.

Example 1. **(continued).** For our B-spline example, the function (1.180) and its dual are displayed in Fig. 1.19. □

Fig. 1.19. Matched primal and dual scaling functions at the cross point.

Wavelets. It can easily be seen that it is not sufficient just to match the non-vanishing wavelets at the cross point since this would yield only one matched wavelet functions, whereas counting the degrees of freedom shows that we have to define two of such functions.

Let us consider the local basis functions on each subdomain

$$\hat{\psi}_e^-(\hat{x}) := \begin{cases} \xi_{j,1}(\hat{x}+1), & e=0, \\ \eta_{j,\nu_{j,M_j}}(\hat{x}+1), & e=1, \end{cases} \qquad \hat{x} \in I_-,$$

$$\hat{\psi}_e^+(\hat{x}) := \begin{cases} \xi_{j,0}(\hat{x}), & e=0, \\ \eta_{j,\nu_{j,1}}(\hat{x}), & e=1, \end{cases} \qquad \hat{x} \in I_+,$$

and let us set (see [23], (5.6))

$$S^0_{j+1}(I_\pm) := \operatorname{span}\{\hat{\psi}^\pm_e : e \in \{0,1\}\}.$$

Any function $v^\pm \in S^0_{j+1}(I_\pm)$ can be written as $v^\pm = \sum_{e\in\{0,1\}} \alpha^\pm_e \hat{\psi}^\pm_e$; let us denote by $\boldsymbol{\alpha}^\pm := (\alpha^\pm_0, \alpha^\pm_1)$ the vectors of the local degrees of freedom. We want to build the local space

$$S^0_{j+1}(I) := \{v \in C^0(\bar{I}) : v_{|I_\pm} \in S^0_{j+1}(I_\pm)\}$$

by matching functions in $S^0_{j+1}(I_-)$ and $S^0_{j+1}(I_+)$, and we want to find a basis for the subspace $W^0_j(I) := \{v \in S^0_{j+1}(I) : (v, \hat{\tilde{\varphi}}_{j,0})_{0,I} = 0\}$. The matching condition between two functions $v^\pm \in S^0_{j+1}(I_\pm)$ reads:

$$\alpha^-_0\, \xi_{j,1}(1) + \alpha^-_1\, \eta_{j,\nu_{j,M_j}}(1) = \alpha^+_0\, \xi_{j,0}(0) + \alpha^+_1\, \eta_{j,\nu_{j,1}}(0),$$

which, in view of the boundary values of the univariate scaling functions and wavelets (see (1.149) on p. 49), is equivalent to

$$\alpha^-_0 + \alpha^-_1 = \alpha^+_0 + \alpha^+_1. \tag{1.181}$$

Next, we enforce the additional condition of orthogonality to $\hat{\tilde{\varphi}}_{j,0}$ which is expressed as

$$\begin{aligned} 0 &= \left(\alpha^-_0 \xi_{j,1}(\cdot+1) + \alpha^-_1 \eta_{j,\nu_{j,M_j}}(\cdot+1), \hat{\tilde{\varphi}}_{j,C_{|[-1,0]}}\right)_{0,[-1,0]} \\ &\quad + \left(\alpha^+_0 \xi_{j,0}(\cdot) + \alpha^+_1 \eta_{j,\nu_{j,1}}(\cdot), \hat{\tilde{\varphi}}_{j,C_{|[0,1]}}\right)_{0,(0,1)} \\ &= \frac{1}{\sqrt{2}}(\alpha^-_0 + \alpha^+_0), \end{aligned} \tag{1.182}$$

where the last equality is a consequence of the biorthogonality on the interval.

We may reformulate condition (1.181) and condition (1.182) multiplied by $\sqrt{2}$ using a matrix-vector notation (as in [23], (5.24)) by

$$\boldsymbol{\mathcal{D}}\boldsymbol{\alpha} = \mathbf{0},$$

where

$$\boldsymbol{\mathcal{D}} := \begin{bmatrix} 1 & 1 & -1 & -1 \\ 1 & 0 & 1 & 0 \end{bmatrix}, \qquad \boldsymbol{\alpha} = (\boldsymbol{\alpha}^-, \boldsymbol{\alpha}^+)^T = (\alpha^-_0, \alpha^-_1, \alpha^+_0, \alpha^+_1)^T.$$

It is easily seen that $\boldsymbol{\mathcal{D}}$ has full rank 2. This implies that $\dim W^0_j(I) = 2$. It is directly seen that

$$\operatorname{Ker} \boldsymbol{\mathcal{D}} = \operatorname{span}\{(0,1,0,1)^T, (1,-1,-1,1)^T\}. \tag{1.183}$$

For the dual system, we have the same condition. So, it remains to find 2 particular choices of $\boldsymbol{\alpha} \in \operatorname{Ker} \boldsymbol{\mathcal{D}}$ and $\tilde{\boldsymbol{\alpha}} \in \operatorname{Ker} \tilde{\boldsymbol{\mathcal{D}}}$, i.e.,

$$\begin{aligned}\boldsymbol{\alpha}^l &:= a_{l,1}\,(0,1,0,1)^T + a_{l,2}\,(1,-1,-1,1)^T,\\ \tilde{\boldsymbol{\alpha}}^l &:= \tilde{a}_{l,1}\,(0,1,0,1)^T + \tilde{a}_{l,2}\,(1,-1,-1,1)^T,\end{aligned} \qquad l=1,2,$$

that will define primal and dual wavelets $\hat{\psi}_j^l$, $\hat{\tilde{\psi}}_j^l$, $l = 1,2$, as

$$\hat{\psi}_j^l(\hat{x}) := \begin{cases} a_{l,2}\,\xi_{j,1}(\hat{x}+1) + (a_{l,1}-a_{l,2})\,\eta_{j,\nu_{j,M_j}}(\hat{x}+1), & \hat{x}\in \bar{I}_-,\\ -a_{l,2}\,\xi_{j,0}(\hat{x}) + (a_{l,1}+a_{l,2})\,\eta_{j,\nu_{j,1}}(\hat{x}), & \hat{x}\in \bar{I}_+.\end{cases} \tag{1.184}$$

The coefficients have to be chosen in order to obtain biorthogonal functions. Using the biorthogonality on the interval, it is readily seen that

$$\delta_{l,m} = (\hat{\psi}_j^l, \hat{\tilde{\psi}}_j^m)_{0,[-1,1]} \;=\; 2\,a_{l,1}\tilde{a}_{m,1} + 4\,a_{l,2}\tilde{a}_{m,2}.$$

This can be rephrased by the matrix equation

$$\boldsymbol{Id} = \boldsymbol{A}\,\boldsymbol{X}\,\tilde{\boldsymbol{A}}^T, \tag{1.185}$$

where

$$\boldsymbol{A} := \begin{bmatrix} a_{1,1} & a_{1,2}\\ a_{2,1} & a_{2,2}\end{bmatrix}, \quad \tilde{\boldsymbol{A}} := \begin{bmatrix} \tilde{a}_{1,1} & \tilde{a}_{1,2}\\ \tilde{a}_{2,1} & \tilde{a}_{2,2}\end{bmatrix}, \quad \boldsymbol{X} := \begin{bmatrix} 2 & 0\\ 0 & 4\end{bmatrix}.$$

Since we have 4 equations for 8 unknowns, one can, in principle, choose 4 coefficients and the remaining 4 are then determined by (1.185).

It is convenient to relabel the wavelets $\hat{\psi}_j^l$, $\hat{\tilde{\psi}}_j^l$, $l = 1,2$, obtained by any particular choice of the coefficients as

$$\hat{\psi}_j^- := \hat{\psi}_j^1, \qquad \hat{\psi}_j^+ := \hat{\psi}_j^2, \tag{1.186}$$

(and similarly for the dual wavelets) so that they are associated in a natural way to the grid points $h_\pm \in \mathcal{H}_j$ located around C and defined as $h_+ := F_+(\nu_{j,1})$, $h_- := F_-(-1+\nu_{j,M_j})$.

The above mentioned freedom in the construction can be used to fulfill additional features, which will now be described.

Additional Features. It can easily be seen that one may choose one wavelet to vanish at the interface. Another option is to choose one wavelet to be symmetric and the other one to be skew-symmetric or to choose all wavelets being reflection invariant around the interface. Note that it is not possible to localize the wavelets in only one subdomain.

Example 1. **(continued).** For our B-spline example, matched wavelets defined in (1.184) with the choice of matrices $\boldsymbol{A}$ and $\tilde{\boldsymbol{A}}$ which guarantee reflection invariance (see above), are shown in Fig. 1.20. □

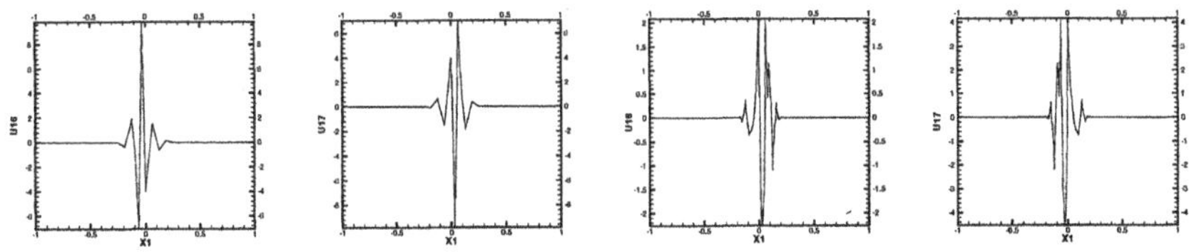

Fig. 1.20. Matched primal and dual wavelets at the cross point.

Another Basis of Matched Functions. Now we aim at defining a basis of the local space $S^0_{j+1}(I)$. Compared to Subsect. 1.5.5 on p. 58, the orthogonality condition (1.182) is missing, so that we obtain the following matching conditions in matrix-vector form

$$\boldsymbol{\mathcal{D}}\,\boldsymbol{\alpha} = \mathbf{0}, \text{ with } \boldsymbol{\mathcal{D}} := (1,1,-1,-1), \quad \boldsymbol{\alpha} = (\boldsymbol{\alpha}^-, \boldsymbol{\alpha}^+)^T = (\alpha_0^-, \alpha_1^-, \alpha_0^+, \alpha_1^+)^T.$$

It is obvious that $\operatorname{Ker}\boldsymbol{\mathcal{C}} = \operatorname{span}\{(1,-1,0,0)^T, (0,0,1,-1)^T, (0,1,0,1)^T\}$ and we have to find 3 particular linear combinations, i.e.,

$$\begin{aligned} \boldsymbol{\alpha}^l &= a_{l,1}\,(1,-1,0,0)^T + a_{l,2}\,(0,0,1,-1)^T + a_{l,3}\,(0,1,0,1)^T, \\ \tilde{\boldsymbol{\alpha}}^l &= \tilde{a}_{l,1}\,(1,-1,0,0)^T + \tilde{a}_{l,2}\,(0,0,1,-1)^T + \tilde{a}_{l,3}\,(0,1,0,1)^T, \end{aligned} \qquad l = 1,2,3,$$

which give rise to the three basis functions

$$\hat{\vartheta}^l_j(\hat{x}) := \begin{cases} a_{l,1}\,\xi_{j,1}(\hat{x}) + (a_{l,3} - a_{l,1})\,\eta_{j,\nu_{j,M_j}}(\hat{x}), & \hat{x} \in I_-, \\ a_{l,2}\,\xi_{j,0}(\hat{x}) + (a_{l,3} - a_{l,2})\,\eta_{j,\nu_{j,1}}(\hat{x}), & \hat{x} \in I_+, \end{cases} \qquad l = 1,2,3, \tag{1.187}$$

In this case, the biorthogonality gives the conditions

$$\begin{aligned} \delta_{l,m} &= (\hat{\vartheta}^l_j, \hat{\tilde{\vartheta}}^m_j)_{L^2(-1,1)} \\ &= 2(a_{l,1}\tilde{a}_{m,1} + a_{l,2}\tilde{a}_{m,2} + a_{l,3}\tilde{a}_{m,3}) \\ &\qquad - a_{l,1}\tilde{a}_{m,3} - a_{l,3}\tilde{a}_{m,1} - a_{l,2}\tilde{a}_{m,3} - a_{l,3}\tilde{a}_{m,2}, \end{aligned}$$

which can be rewritten as $\boldsymbol{Id} = \boldsymbol{B}\,\boldsymbol{Y}\,\tilde{\boldsymbol{B}}^T$, where

$$\boldsymbol{B} := \begin{bmatrix} a_{1,1} & a_{1,2} & a_{1,3} \\ a_{2,1} & a_{2,2} & a_{2,3} \\ a_{3,1} & a_{3,2} & a_{3,3} \end{bmatrix}, \quad \tilde{\boldsymbol{B}} := \begin{bmatrix} \tilde{a}_{1,1} & \tilde{a}_{1,2} & \tilde{a}_{1,3} \\ \tilde{a}_{2,1} & \tilde{a}_{2,2} & \tilde{a}_{2,3} \\ \tilde{a}_{3,1} & \tilde{a}_{3,2} & \tilde{a}_{3,3} \end{bmatrix},$$

and

$$\boldsymbol{Y} := \begin{bmatrix} 2 & 0 & -1 \\ 0 & 2 & -1 \\ -1 & -1 & 2 \end{bmatrix}.$$

After choosing one particular solution of this algebraic system, we relabel the functions as

$$\hat{\vartheta}^-_j := \hat{\vartheta}^1_j, \quad \hat{\vartheta}^0_j := \hat{\vartheta}^2_j, \quad \hat{\vartheta}^+_j := \hat{\vartheta}^3_j; \tag{1.188}$$

the dual functions are defined similarly.

Additional Features. One obtains the same features as in the above case with the difference that it is possible to construct three functions such that only one of them is localized in both subdomains. The remaining two are supported in only one subdomain.

Example 1. **(continued).** We give one particular example of three functions, one located in I_-, one in I_+ and one in both subintervals. For the latter one, also the vanishing moment property is preserved. Let

$$\boldsymbol{B} := \begin{bmatrix} 1\,0\,0 \\ 0\,0\,1 \\ 0\,1\,0 \end{bmatrix}. \tag{1.189}$$

It is readily seen that $\boldsymbol{B}^{-1} = \boldsymbol{B}^T$, hence we obtain the coefficients for the dual functions by

$$\tilde{\boldsymbol{B}} = \boldsymbol{B}\boldsymbol{Y}^{-t} = \begin{bmatrix} 3/4 & 1/4 & 1/2 \\ 1/2 & 1/2 & 1 \\ 1/4 & 3/4 & 1/2 \end{bmatrix}. \tag{1.190}$$

These particular functions are displayed in Fig. 1.21. □

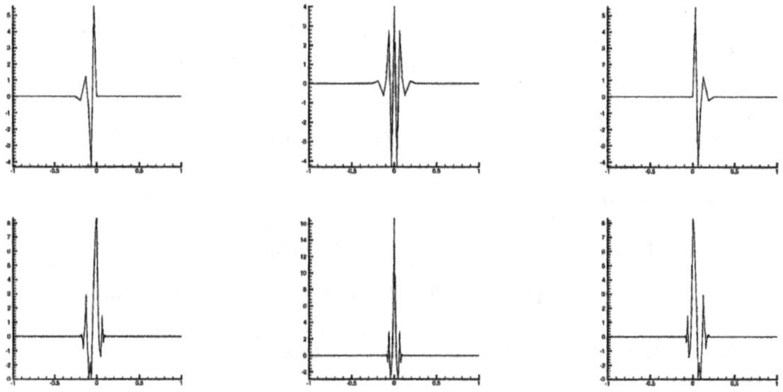

Fig. 1.21. Primal and dual basis functions defined by (1.189) and (1.190).

1.5.6 Bivariate Matched Wavelets

We shall now construct matched two-dimensional wavelets, by firstly considering an interior cross point, next a boundary cross point and finally the common side of two subdomains. In each case, we indicate how wavelets can be defined, which have the most localized support.

Matched Wavelets Around an Interior Cross Point. We describe the construction of wavelets associated to grid points $h \in \mathcal{H}_j$ which are close to a cross point C, at which N_C subdomains meet, see, e.g. Fig. 1.22. We assume that these subdomains are (re-)labeled by $\Omega_1, \ldots, \Omega_{N_C}$, in a counterclockwise order. Moreover, we set $\Gamma_{i,i+1} := \partial\Omega_i \cap \partial\Omega_{i+1}$. We start with the situation in which C is interior to Ω. In this case, it is convenient to set $\Omega_{N_C+1} := \Omega_1$.

At first, we deal with a particular choice of the mappings to the reference domain. Next, we shall show that all other possibilities can be easily reduced to this choice, which – therefore – can be thought of as a *reference situation*.

So, suppose that, for all $i \in \{1, \ldots, N_C\}$, one has $C = F_i(0,0)$ and $\Gamma_{i,i+1} = F_i(\hat{\sigma}_{01}) = F_{i+1}(\hat{\sigma}_{10})$, where $\hat{\sigma}_{01} := \{(0, \hat{x}_2) : 0 \le \hat{x}_2 \le 1\}$ and $\hat{\sigma}_{10} := \{(\hat{x}_1, 0) : 0 \le \hat{x}_1 \le 1\}$. The grid points surrounding C, to which we will associate the matched wavelets, are the $2N_C$ points $h_{C,l}$ defined as follows:

$$h_{C,2i-1} = F_i(\nu_{j,1}, \nu_{j,1}), \qquad h_{C,2i} = F_i(0, \nu_{j,1}), \qquad 1 \le i \le N_C. \tag{1.191}$$

Note that each evenly numbered point belongs to a side meeting at C, whereas each oddly numbered point is internal to a subdomain meeting at C (see Fig. 1.22).

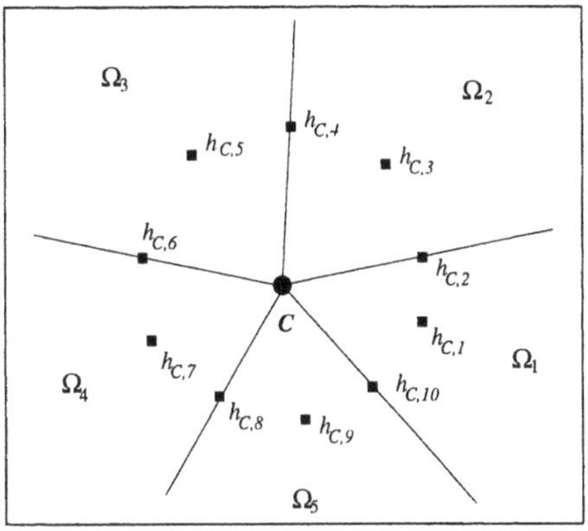

Fig. 1.22. Grid points $h_{C,i}$, $i = 1, \ldots, 10$, around a cross point C common to 5 subdomains.

Dropping the index j, let us set

$$\begin{aligned}
\psi_{00}^{(i)}(x) &= \hat{\psi}_{00}(\hat{x}) = \xi_{j,0}(\hat{x}_1)\, \xi_{j,0}(\hat{x}_2), \\
\psi_{01}^{(i)}(x) &= \hat{\psi}_{01}(\hat{x}) = \xi_{j,0}(\hat{x}_1)\, \eta_{j,\nu_{j,1}}(\hat{x}_2), \\
\psi_{10}^{(i)}(x) &= \hat{\psi}_{10}(\hat{x}) = \eta_{j,\nu_{j,1}}(\hat{x}_1)\, \xi_{j,0}(\hat{x}_2), \\
\psi_{11}^{(i)}(x) &= \hat{\psi}_{11}(\hat{x}) = \eta_{j,\nu_{j,1}}(\hat{x}_1)\, \eta_{j,\nu_{j,1}}(\hat{x}_2),
\end{aligned}$$

(see [23], formula (5.5)), as well as $S^C_{j+1}(\Omega_i) := \text{span}\{\psi^{(i)}_e : e \in E^2\}$. A function $v^{(i)} \in S^C_{j+1}(\Omega_i)$ is written as

$$v^{(i)} = \sum_{e \in E^2} \alpha^{(i)}_e \psi^{(i)}_e;$$

we shall introduce the column vector $\boldsymbol{\alpha}^{(i)} := (\alpha^{(i)}_e)_{e \in E^2}$. In order to characterize the local space

$$S^C_{j+1}(\Omega) := \{v \in C^0(\bar{\Omega}) : v_{|\Omega_i} \in S^C_{j+1}(\Omega_i) \text{ if } i \in \{1, \dots, N_C\},\\ v_{|\Omega_i} \equiv 0 \text{ elsewhere}\},$$

we proceed as in [23] (see Sect. 5.2), i.e., we enforce the continuity among the $v^{(i)}$ by considering the point C firstly, and the sides $\Gamma_{i,i+1}$, secondly. Recalling (1.149) on p. 49, the continuity at C yields the set of linearly independent conditions

$$\alpha^{(i)}_{00} + \alpha^{(i)}_{01} + \alpha^{(i)}_{10} + \alpha^{(i)}_{11} = \alpha^{(i+1)}_{00} + \alpha^{(i+1)}_{01} + \alpha^{(i+1)}_{10} + \alpha^{(i+1)}_{11}, \qquad 1 \le i \le N_C - 1, \tag{1.192}$$

or

$$\boldsymbol{c}_0 \cdot \boldsymbol{\alpha}^{(i)} = \boldsymbol{c}_0 \cdot \boldsymbol{\alpha}^{(i+1)}, \qquad 1 \le i \le N_C - 1,$$

with $\boldsymbol{c}_0 = (1, 1, 1, 1)$. Denoting by $\boldsymbol{\alpha} := (\boldsymbol{\alpha}^{(i)})_{i=1,\dots,N_C} \in \mathbb{R}^{4N_C}$ the column vector of all degrees of freedom, these conditions can be written in matrix-vector form as

$$\boldsymbol{C}_0 \, \boldsymbol{\alpha} = \boldsymbol{0}, \qquad \text{where} \qquad \boldsymbol{C}_0 = \begin{bmatrix} \boldsymbol{c}_0 & -\boldsymbol{c}_0 & & \\ & \ddots & \ddots & \\ & & \boldsymbol{c}_0 & -\boldsymbol{c}_0 \end{bmatrix} \in \mathbb{R}^{(N_C-1)\times 4N_C}. \tag{1.193}$$

Let us now enforce continuity along the sides $\Gamma_{i,i+1}$. To this end, observe that

$$v^{(i)}_{|\Gamma_{i,i+1}}(x) = \lambda_j\Big((\alpha^{(i)}_{00} + \alpha^{(i)}_{10})\xi_{j,0}(\hat{x}_2) + (\alpha^{(i)}_{01} + \alpha^{(i)}_{11})\eta_{j,\nu_{j,1}}(\hat{x}_2)\Big),\\ v^{(i+1)}_{|\Gamma_{i,i+1}}(x) = \lambda_j\Big((\alpha^{(i+1)}_{00} + \alpha^{(i+1)}_{01})\xi_{j,0}(\hat{x}_1) + (\alpha^{(i+1)}_{10} + \alpha^{(i+1)}_{11})\eta_{j,\nu_{j,1}}(\hat{x}_1)\Big), \tag{1.194}$$

with λ_j defined in (1.150) on p. 49. Because of the linear independence of the univariate functions, the matching is equivalent to

$$\alpha^{(i)}_{00} + \alpha^{(i)}_{10} = \alpha^{(i+1)}_{00} + \alpha^{(i+1)}_{01},\\ \alpha^{(i)}_{01} + \alpha^{(i)}_{11} = \alpha^{(i+1)}_{10} + \alpha^{(i+1)}_{11}. \tag{1.195}$$

Since we have already enforced the continuity at $C \in \Gamma_{i,i+1}$ (see (1.192)), it is enough to require that a particular linear combination of the latter equations holds; precisely, we enforce

$$\alpha^{(i)}_{00} - \alpha^{(i)}_{01} + \alpha^{(i)}_{10} - \alpha^{(i)}_{11} = \alpha^{(i+1)}_{00} + \alpha^{(i+1)}_{01} - \alpha^{(i+1)}_{10} - \alpha^{(i+1)}_{11}$$

(for details, see [23], Proposition 5.2).

Introducing the vectors $\boldsymbol{c}' = (1,-1,1,-1)$ and $\boldsymbol{c}'' = (1,1,-1,-1)$, these conditions can be rephrased as

$$\mathcal{C}_1\, \boldsymbol{\alpha} = \boldsymbol{0}, \qquad \text{where} \qquad \mathcal{C}_1 = \begin{bmatrix} \boldsymbol{c}' & -\boldsymbol{c}'' & & \\ & \ddots & \ddots & \\ & & \boldsymbol{c}' & -\boldsymbol{c}'' \\ -\boldsymbol{c}'' & & & \boldsymbol{c}' \end{bmatrix} \in \mathbb{R}^{N_C \times 4N_C}. \tag{1.196}$$

We are interested in finding a basis for the subspace $W_j^C(\Omega) := \{v \in S_{j+1}^C(\Omega) : \langle v, \tilde{\varphi}_{j,C} \rangle_\Omega = 0\}$. Recalling that, by (1.176) on p. 55,

$$\tilde{\varphi}_{j,C|\Omega_i}(x) = \frac{1}{\sqrt{N_C}}\,(\tilde{\xi}_{j,0} \otimes \tilde{\xi}_{j,0})(\hat{x}), \quad \hat{x} = G_i(x), \; i = 1, \ldots, N_C,$$

we obtain the condition

$$\sum_{i=1}^{N_C} \alpha_{00}^{(i)} = 0.$$

So, introducing the vector $\boldsymbol{b} := (1,0,0,0)$, all conditions enforced so far can be summarized in

$$\mathcal{D}\, \boldsymbol{\alpha} = \boldsymbol{0}, \qquad \text{where} \quad \mathcal{D} = \begin{bmatrix} \mathcal{C}_0 \\ \mathcal{C}_1 \\ \mathcal{B} \end{bmatrix} \in \mathbb{R}^{2N_C \times 4N_C}, \quad \mathcal{B} := (\boldsymbol{b}, \ldots, \boldsymbol{b}) \in \mathbb{R}^{1 \times 4N_C}. \tag{1.197}$$

It can easily be seen that

$$\mathcal{D}\mathcal{D}^T = \left[\begin{array}{ccccc|ccc|c} 8 & -4 & & & 0 & & & & \\ -4 & 8 & -4 & & & & & & \\ & \ddots & \ddots & \ddots & & & & & \\ & & \ddots & \ddots & -4 & & & & \\ 0 & & & -4 & 8 & & & & \\ \hline & & & & & 8 & & & \\ & & & & & & \ddots & & \\ & & & & & & & 8 & \\ \hline & & & & & & & & N_C \end{array}\right] \in \mathbb{R}^{2N_C \times 2N_C};$$

the evident symmetric positive definite character of $\mathcal{D}\mathcal{D}^T$ means that $\mathcal{D}$ has full rank. This implies the existence of exactly $2N_C$ linearly independent functions in $W_j^C(\Omega)$, i.e., $\dim W_j^C(\Omega) = 2N_C$. The parallel construction for $\tilde{W}_j^C(\Omega)$ leads to the condition $\mathcal{D}\,\tilde{\boldsymbol{\alpha}} = \boldsymbol{0}$ (note that the matrix $\mathcal{D}$ is the same as for the primal system). Finally, we enforce biorthogonality between the primal and dual basis functions determined in this way. To this end, we choose a basis $\{\boldsymbol{\kappa}_1, \ldots, \boldsymbol{\kappa}_{2N_C}\}$ in Ker $\boldsymbol{\mathcal{D}}$, and we look for linear combinations of these vectors

$$\boldsymbol{\alpha}^l = \sum_{m=1}^{2N_C} a_{l,m}\, \boldsymbol{\kappa}_m, \qquad \tilde{\boldsymbol{\alpha}}^l = \sum_{m=1}^{2N_C} \tilde{a}_{l,m}\, \boldsymbol{\kappa}_m, \qquad 1 \le l \le 2N_C,$$

such that, if $\boldsymbol{\alpha}^l$ is decomposed as $(\boldsymbol{\alpha}^{l,(i)})_{i=1,\ldots,2N_C}$, the corresponding wavelets

$$\psi^l_{j,C}(x) := \begin{cases} \sum_{e\in E^2} \alpha_e^{l,(i)}\, \hat{\psi}_e(\hat{x}), & \text{if } x \in \Omega_i,\ 1 \le i \le N_C, \\ 0, & \text{elsewhere,} \end{cases}$$

and the dual ones $\tilde{\psi}^l_{j,C}$ defined in a similar manner, form a biorthogonal system. Setting $\boldsymbol{\mathcal{K}} := (\boldsymbol{\kappa}_m)_{m=1,\ldots,2N_C}$, $\boldsymbol{A} := (a_{l,m})_{l,m=1,\ldots,2N_C}$, $\tilde{\boldsymbol{A}} := (\tilde{a}_{l,m})_{l,m=1,\ldots,2N_C}$, and $\boldsymbol{\mathcal{A}} := (\boldsymbol{\alpha}^l)_{l=1,\ldots,2N_C} = \boldsymbol{\mathcal{K}}\boldsymbol{A}^T$, $\tilde{\boldsymbol{\mathcal{A}}} := (\tilde{\boldsymbol{\alpha}}^l)_{l=1,\ldots,2N_C} = \boldsymbol{\mathcal{K}}\tilde{\boldsymbol{A}}^T$, and exploiting the biorthogonality property in each subdomain, we can express the biorthogonality condition in the form $\boldsymbol{\mathcal{A}}^T\,\tilde{\boldsymbol{\mathcal{A}}} = \boldsymbol{Id}$, i.e.,

$$\boldsymbol{A}\boldsymbol{\mathcal{K}}^T\boldsymbol{\mathcal{K}}\tilde{\boldsymbol{A}}^T = \boldsymbol{Id}. \tag{1.198}$$

Since $\boldsymbol{\mathcal{K}}$ obviously has full rank, $\boldsymbol{\mathcal{K}}^T\boldsymbol{\mathcal{K}}$ is regular, so this matrix equation has a solution. In particular, one can choose $\boldsymbol{A} = \boldsymbol{Id}$ and consequently $\tilde{\boldsymbol{A}} = (\boldsymbol{\mathcal{K}}^T\boldsymbol{\mathcal{K}})^{-1}$.

Once the biorthogonal wavelets $\psi^l_{j,C}$ and $\tilde{\psi}^l_{j,C}$ $(1 \le l \le 2N_C)$ have been determined, they can be associated to the $2N_C$ grid points $h_{C,l}$ surrounding C, defined in (1.191).

Reduction to the Reference Situation. Let us now show that we can reduce any interior cross point situation to the one described above. To this end, let us consider any subdomain Ω_i having C as a vertex. Then, we have the following 4 cases:
a) $C = F_i(0,0)$, b) $C = F_i(1,1)$,
c) $C = F_i(0,1)$, d) $C = F_i(1,0)$.
Recalling the assumption $\det(JF_i) > 0$ in (1.159) on p. 52, it follows that in cases a) and b) the indices of the frozen coordinates of $\hat{\Gamma}_{i-1,i}$ and $\hat{\Gamma}_{i,i+1}$ are given by $\mathcal{L}_{\hat{\Gamma}_{i-1,i}} = \{2\}$, $\mathcal{L}_{\hat{\Gamma}_{i,i+1}} = \{1\}$, whereas in cases c) and d) one has $\mathcal{L}_{\hat{\Gamma}_{i-1,i}} = \{1\}$, $\mathcal{L}_{\hat{\Gamma}_{i,i+1}} = \{2\}$. It is straightforward to see that the matching conditions along $\Gamma_{i-1,i}$ and $\Gamma_{i,i+1}$ in cases a) and b) yield the same vectors $\boldsymbol{c}'$ and $\boldsymbol{c}''$ defined above, whereas the roles of these vectors are interchanged in the remaining cases.

If $\boldsymbol{\alpha}^{(i)} = (\alpha^{(i)}_{00}, \alpha^{(i)}_{01}, \alpha^{(i)}_{10}, \alpha^{(i)}_{11})^T$, denote by $\check{\boldsymbol{\alpha}}^{(i)} := (\alpha^{(i)}_{00}, \alpha^{(i)}_{10}, \alpha^{(i)}_{01}, \alpha^{(i)}_{11})^T$ the modified vector and by $\check{\boldsymbol{\mathcal{D}}}$ the matrix obtained by modifying $\boldsymbol{\mathcal{D}}$ according to cases c) and d). Since it is readily seen that $\boldsymbol{c}_0 \cdot (\boldsymbol{c}')^T = \boldsymbol{c}_0 \cdot (\boldsymbol{c}'')^T = 0$, $\boldsymbol{b} \cdot (\boldsymbol{c}')^T = \boldsymbol{b} \cdot (\boldsymbol{c}'')^T = 0$, $\boldsymbol{c}' \cdot \boldsymbol{\alpha}^{(i)} = \boldsymbol{c}'' \cdot \check{\boldsymbol{\alpha}}^{(i)}$ as well as $\boldsymbol{c}'' \cdot \boldsymbol{\alpha}^{(i)} = \boldsymbol{c}' \cdot \check{\boldsymbol{\alpha}}^{(i)}$, we obtain, as desired, $\boldsymbol{\mathcal{D}}\,\boldsymbol{\alpha} = \check{\boldsymbol{\mathcal{D}}}\,\check{\boldsymbol{\alpha}}$.

This procedure can be applied to all subdomains meeting at C, and so we are back to the *reference situation.*

The biorthogonalization is performed following the same guidelines described above; obviously, the definition of the wavelets and the associated grid points has to be adapted to the specific orientation of the mappings to the reference domain.

Wavelets with Minimal Support. It can readily be seen that it is impossible to have a wavelet supported in only one subdomain. One obtains three different types of wavelets. The first two ones are supported in two neighbored subdomains. One can show that all but one functions can be of that type. The remaining one wavelet is supported in all subdomains that meet at the crosspoint, [24].

Example 1. **(continued).** We consider the situation in which four subdomains $\Omega_1, \ldots, \Omega_4$ meet at C (for simplicity, we assume linear parametric mappings from each subdomain to the reference domain). For our B-spline example, we show in Fig. 1.23 on p. 68 the matched scaling function and the three different types of wavelets associated to grid points around C. □

Tensor Products of Matched Univariate Functions. A common situation for an internal cross point is the case $N_C = 4$, i.e., four subdomains meeting at C. In such a geometry, it is easy to construct a basis for $W_j^C(\Omega)$ by properly tensorising the univariate matched functions defined in Sect. 1.5.5 on p. 56.

First of all, let us note that, by possibly introducing appropriate parametric mappings, we can reduce ourselves to the situation in which each subdomain is the image of one of the subdomains $I_\pm \times I_\pm$ (we use here the notation set at the beginning of Sect. 1.5.5), and C is the image of $\hat{C} = (0,0)$. Then, let us consider the set of univariate functions given by the scaling function $\hat{\varphi}_{j,0}$ defined in (1.180), the wavelets $\hat{\psi}_j^l$ defined in (1.184)-(1.186) on p. 60 and the functions $\hat{\vartheta}_j^l$ defined in (1.187)-(1.188) on p. 61. A basis in $W_j^C(\Omega)$ is obtained by taking the image of 8 linearly independent tensor products of such functions satisfying the condition of orthogonality to the dual scaling function $\hat{\tilde{\varphi}}_{j,0} = \hat{\tilde{\varphi}}_{j,0} \otimes \hat{\tilde{\varphi}}_{j,0}$. For example, a possible choice is

$$
\begin{array}{ccc}
\hat{\psi}_j^- \otimes \hat{\vartheta}_j^+, & \hat{\varphi}_{j,0} \otimes \hat{\psi}_j^+, & \hat{\psi}_j^+ \otimes \hat{\vartheta}_j^+, \\
\hat{\psi}_j^- \otimes \hat{\vartheta}_j^0, & & \hat{\psi}_j^+ \otimes \hat{\vartheta}_j^0, \\
\hat{\psi}_j^- \otimes \hat{\vartheta}_j^-, & \hat{\varphi}_{j,0} \otimes \hat{\psi}_j^-, & \hat{\psi}_j^+ \otimes \hat{\vartheta}_j^-,
\end{array}
$$

(obviously, these functions are extended by zero outside the union of the four subdomains) whose association to the 8 wavelet grid points around C is self evident. Note that this construction does not necessarily require the functions $\hat{\vartheta}_j^l$ to have minimal support, although efficiency will be enhanced by this feature.

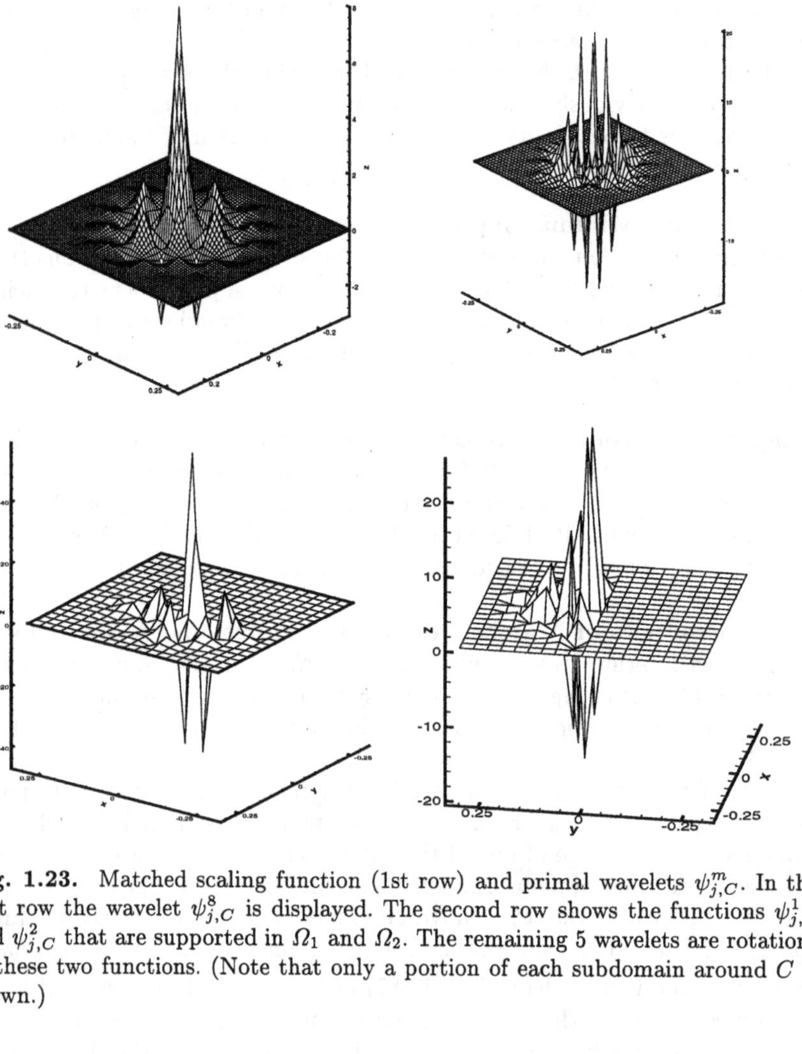

Fig. 1.23. Matched scaling function (1st row) and primal wavelets $\psi_{j,C}^m$. In the first row the wavelet $\psi_{j,C}^8$ is displayed. The second row shows the functions $\psi_{j,C}^1$ and $\psi_{j,C}^2$ that are supported in Ω_1 and Ω_2. The remaining 5 wavelets are rotations of these two functions. (Note that only a portion of each subdomain around C is shown.)

Matched Wavelets Around a Boundary Cross Point. Let us now consider the situation in which $C \in \partial\Omega$ is common to N_C subdomains $\Omega_1, \ldots, \Omega_{N_C}$ ordered counterclockwise. We assume that Ω_1 (Ω_{N_C}, resp.) has a side, termed Γ_1 (Γ_{N_C}, resp.), which contains C and lies on $\partial\Omega$. Then, the following cases may occur:

a) $\Gamma_1, \Gamma_{N_C} \in \Gamma_{\mathrm{Neu}}$

b) $\Gamma_1 \in \Gamma_{\mathrm{Dir}}, \Gamma_{N_C} \in \Gamma_{\mathrm{Neu}}$ c) $\Gamma_1 \in \Gamma_{\mathrm{Neu}}, \Gamma_{N_C} \in \Gamma_{\mathrm{Dir}}$

d) $\Gamma_1, \Gamma_{N_C} \in \Gamma_{\mathrm{Dir}}$.

Since all arguments concerning biorthogonalization carry over from the interior cross point case, in the sequel we will only detail the matching and orthogonality conditions for each case separately.

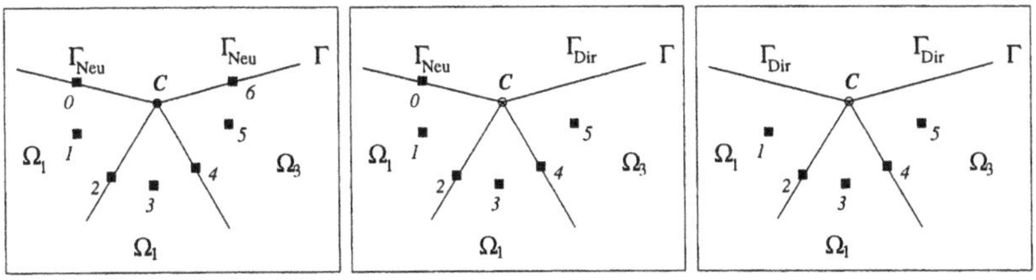

Fig. 1.24. Grid points $h_{C,i}$ (labeled only by i) around a boundary cross point in the 3 different cases (from left to right: pure Neumann, mixed and pure Dirichlet case). Note that the cross point C does only correspond to a scaling function in the pure Neumann case.

Pure Neumann Case. Let us start by considering case a). The matching at the cross point C only differs from the interior cross point case by the absence of a matching condition between Ω_1 and Ω_{N_C}. However, in the interior case, this condition turned out to be linearly dependent on the other ones, hence, there was no need to explicitly enforce it. This implies that the matrices describing the matching at C are the same: $\boldsymbol{C}_0^{\text{Neu}} = \boldsymbol{C}_0$, see (1.193).

The matching conditions along the sides are the same as well, with the only difference that now the last row in (1.196) is missing:

$$\boldsymbol{C}_1^{\text{Neu}} = \begin{bmatrix} \boldsymbol{c}' & -\boldsymbol{c}'' & & \\ & \ddots & \ddots & \\ & & \boldsymbol{c}' & -\boldsymbol{c}'' \end{bmatrix} \in \mathbb{R}^{(N_C-1)\times 4N_C}.$$

Since $C \in \Gamma_{\text{Neu}}$, we observe that there exists a dual scaling function $\tilde{\varphi}_{j,C}$ associated to C. Consequently, we have to enforce orthogonality to this function, so we end up with the set of conditions

$$\boldsymbol{\mathcal{D}}^{\text{Neu}}\,\boldsymbol{\alpha} = \boldsymbol{0}, \qquad \text{with} \quad \boldsymbol{\mathcal{D}}^{\text{Neu}} = \begin{bmatrix} \boldsymbol{C}_0 \\ \boldsymbol{C}_1^{\text{Neu}} \\ \boldsymbol{\mathcal{B}} \end{bmatrix} \in \mathbb{R}^{(2N_C-1)\times 4N_C},$$

where $\boldsymbol{\mathcal{B}}$ is defined in (1.197). As above, it is easily seen that $\boldsymbol{\mathcal{D}}^{\text{Neu}}$ has full rank; this implies that $\dim W_j^C(\Omega) = 2N_C + 1$. This is precisely the number of grid points surrounding C to which these wavelets can be associated; in the reference situation, they are the points (1.191) on p. 63 and the point $h_{C,0} := F_1(\nu_{j,1}, 0)$ (see Fig. 1.24 on p. 69, left).

Mixed Neumann/Dirichlet Case. Obviously, the mixed Neumann/Dirichlet cases b) and c) can be viewed as symmetric ones, so we will only detail case b) here. Since $C \in \bar{\Gamma}_{\text{Dir}}$, all functions in $S^C_{j+1}(\Omega)$ have to vanish at C. This means that we have to add one more condition to those posed at C, i.e., we obtain the matrix

$$\boldsymbol{C}_0^{\text{Mix}} = \begin{bmatrix} -\boldsymbol{c}_0 & & & \\ \boldsymbol{c}_0 & -\boldsymbol{c}_0 & & \\ & \ddots & \ddots & \\ & & \boldsymbol{c}_0 & -\boldsymbol{c}_0 \end{bmatrix} \in \mathbb{R}^{N_C \times 4N_C}.$$

Furthermore, again we assume to be in the reference situation described above, which implies that $\mathcal{L}_{\hat{\Gamma}_1} = \{2\}$ and $\mathcal{L}_{\hat{\Gamma}_{1,2}} = \{1\}$ for the domain Ω_1, i.e., the second coordinate of $\hat{\Gamma}_1$ is frozen. Let us consider a function $v^{(1)} \in S^C_{j+1}(\Omega_1)$, which is written as

$$v^{(1)}(x) = \sum_{e \in E^2} \alpha_e^{(1)} \hat{\psi}_e(\hat{x}), \qquad x \in \Omega_1.$$

Observing that

$$v^{(1)}_{|\Gamma_1}(x) = \lambda_j \left((\alpha_{00}^{(1)} + \alpha_{01}^{(1)})\, \xi_{j,0}(\hat{x}_1) + (\alpha_{10}^{(1)} + \alpha_{11}^{(1)})\, \eta_{j,\nu_{j,1}}(\hat{x}_1) \right),$$

we enforce $v^{(1)}_{|\Gamma_1} \equiv 0$ if and only if the relations $\alpha_{00}^{(1)} + \alpha_{01}^{(1)} = 0$ and $\alpha_{10}^{(1)} + \alpha_{11}^{(1)} = 0$ are satisfied. In other words, $v^{(1)}$ has to be written as

$$v^{(1)}(x) = \sqrt{2} \left(\alpha_{01}^{(1)} \xi_{j,0}(\hat{x}_1) + \alpha_{11}^{(1)} \eta_{j,\nu_{j,1}}(\hat{x}_1) \right) \eta^D_{j,\nu_{j,1}}(\hat{x}_2),$$

where $\eta^D_{j,\nu_{j,1}}$ is the univariate wavelet vanishing at 0, defined in (1.155) on p. 50. Since $v^{(1)}$ has already been set to 0 at C, we now enforce the linear combination

$$-(\alpha_{00}^{(1)} + \alpha_{01}^{(1)}) + (\alpha_{10}^{(1)} + \alpha_{11}^{(1)}) = 0.$$

The matrix containing the matching and boundary conditions along the sides takes the form

$$\boldsymbol{C}_1^{\text{Mix}} = \begin{bmatrix} -\boldsymbol{c}'' & & & & \\ \boldsymbol{c}' & -\boldsymbol{c}'' & & & \\ & \ddots & \ddots & & \\ & & \boldsymbol{c}' & -\boldsymbol{c}'' & \\ & & & \boldsymbol{c}' & -\boldsymbol{c}'' \end{bmatrix} \in \mathbb{R}^{N_C \times 4N_C}.$$

Since there is no scaling function associated to C, we end up with the system

$$\boldsymbol{\mathcal{D}}^{\text{Mix}} \boldsymbol{\alpha} = \boldsymbol{0}, \quad \text{with} \quad \boldsymbol{\mathcal{D}}^{\text{Mix}} = \begin{bmatrix} \boldsymbol{C}_0^{\text{Mix}} \\ \boldsymbol{C}_1^{\text{Mix}} \end{bmatrix} \in \mathbb{R}^{2N_C \times 4N_C}.$$

Again, $\mathcal{D}^{\mathrm{Mix}}$ is easily shown to have full rank, so that $\dim W_j^C(\Omega) = 2N_C$; this is precisely the number of grid points surrounding C, to which the wavelets are associated (see Fig. 1.24 on p. 69, center).

Pure Dirichlet Case. In case d), the matching and boundary conditions at C obviously coincide with those of cases b) and c). As far as the conditions at the sides are concerned, working out as before we end up with the matrix

$$\mathcal{C}_1^{\mathrm{Dir}} = \begin{bmatrix} -\boldsymbol{c}'' & & & & \\ \boldsymbol{c}' & -\boldsymbol{c}'' & & & \\ & & \ddots & \ddots & \\ & & & \boldsymbol{c}' & -\boldsymbol{c}'' \\ & & & & \boldsymbol{c}' \end{bmatrix} \in \mathbb{R}^{(N_C+1)\times 4N_C}$$

and the whole system takes the form

$$\mathcal{D}^{\mathrm{Dir}}\,\boldsymbol{\alpha} = \mathbf{0}, \qquad \text{with} \quad \mathcal{D}^{\mathrm{Dir}} = \begin{bmatrix} \mathcal{C}_0^{\mathrm{Mix}} \\ \mathcal{C}_1^{\mathrm{Dir}} \end{bmatrix} \in \mathbb{R}^{(2N_C+1)\times 4N_C}.$$

Once again, $\mathcal{D}^{\mathrm{Dir}}$ has full rank, so that $\dim W_j^C(\Omega) = 2N_C - 1$ (see Fig. 1.24 on p. 69, right).

Wavelets with Minimal Support. As opposed to the interior crosspoint case it is sometimes possible to construct certain wavelets that are supported in only one subdomain. In the pure Neumann case one can construct $2N_C+1$ linearly independent functions, two of them supported in exactly one subdomain, one supported in all subdomains matching at C, while the remaining ones are supported across two consecutive subdomains.

In the pure Dirichlet case no global wavelet is needed. In fact, it is possible to have one function supported in one subdomain and the remaining ones in two subdomains.

For the mixed Dirichlet/Neumann case one can use the functions constructed for the pure Dirichlet case and simply add one function which is localized in one subdomain.

Example 1. **(continued).** We consider an L-shaped domain made up by 3 square subdomains meeting at C. We enforce homogeneous Dirichlet conditions on the whole boundary. The two different types of wavelets produced by our construction are displayed in Fig. 1.25 on p. 72. □

Matched Wavelets Across a Side. Let us consider two subdomains Ω_+ and Ω_- having a common side $\sigma := \bar{\Omega}_+ \cap \bar{\Omega}_-$, see Fig. 1.26 on p. 72. Moreover, let us denote by A and B the two endpoints of σ. As in the cross point case, we may reduce ourselves to a reference situation. Instead of thinking each subdomain as the image of the reference domain $\hat{\Omega}$, here it is natural to

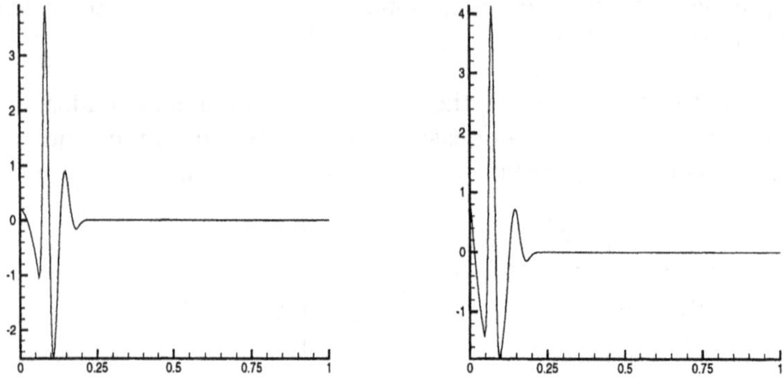

Fig. 1.25. Wavelets $\psi^2_{j,C}$ and $\psi^3_{j,C}$ around a Dirichlet boundary cross point. (Note again that only a portion of each subdomain around C is shown.)

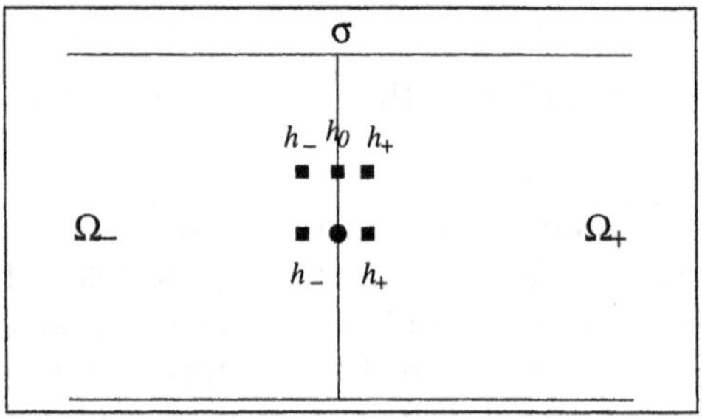

Fig. 1.26. Gridpoints for the matching across a side for two subdomains Ω_+ and Ω_- having a common side $\sigma := \bar{\Omega}_+ \cap \bar{\Omega}_-$

think Ω_- (Ω_+, resp.) as the image of the domain $I_- \times I_+$ ($I_+ \times I_+$, resp.), with

$$A = F_-(0,0) = F_+(0,0), \qquad B = F_-(0,1) = F_+(0,1). \tag{1.199}$$

Since the reasoning for the reduction to the reference situation is analogous to the cross point case, we drop these arguments here. Now, (1.199) implies that the set of grid points on σ is given by

$$\begin{aligned} \mathcal{H}_\sigma &:= \{h \,:\, h = F_+(0, \hat{h}_2),\ \hat{h}_2 \in \mathcal{I}_j \cup \mathcal{J}_j\} \\ &= \{h \,:\, h = F_-(0, \hat{h}_2),\ \hat{h}_2 \in \mathcal{I}_j \cup \mathcal{J}_j\}. \end{aligned}$$

The grid points $F_+(0, \nu_{j,1})$ and $F_+(0, \nu_{j,M_j})$ are already associated to wavelets $\psi_{j,h}$, since these points correspond to the cross points A and B, respectively. Hence, we are left with the points

$$\mathcal{H}_\sigma^{int} := \mathcal{H}_\sigma \setminus \{F_+(0,\nu_{j,1}), F_+(0,\nu_{j,M_j})\}.$$

Let us first consider the case $h \in \mathcal{H}_\sigma^{int}$ with $\hat{h}_2 \in \mathcal{I}_j$. Consequently, there exists a scaling function

$$\varphi_{j,h}(x) := \frac{1}{\sqrt{2}}\xi_{j,\hat{h}_2}(\hat{x}_2) \times \begin{cases} \xi_{j,1}(\hat{x}_- + 1), & \text{if } x = F_-(\hat{x}_-, \hat{x}_2) \in \Omega_-, \\ \xi_{j,0}(\hat{x}_+), & \text{if } x = F_+(\hat{x}_+, \hat{x}_2) \in \Omega_+, \\ 0, & \text{elsewhere,} \end{cases}$$

associated to h. The basis functions of the local spaces $S_{j+1}^h(\Omega_\pm)$ are then given by

$$\psi_e^-(x) := \xi_{j,\hat{h}_2}(\hat{x}_2) \times \begin{cases} \xi_{j,1}(\hat{x}_- + 1), & e = 0, \\ \eta_{j,\nu_{j,M_j}}(\hat{x}_- + 1), & e = 1, \end{cases}$$

and

$$\psi_e^+(x) := \xi_{j,\hat{h}_2}(\hat{x}_2) \times \begin{cases} \xi_{j,0}(\hat{x}_+), & e = 0, \\ \eta_{j,\nu_{j,1}}(\hat{x}_+), & e = 1. \end{cases}$$

This shows that the matching at h is equivalent to the matching at a univariate interface point. Considering the wavelet functions $\hat{\psi}_j^l$ defined in (1.186) for $l = -, +$, we end up with the two wavelets

$$\psi_{j,h_\pm}(x) = \begin{cases} (\hat{\psi}_j^\pm \otimes \xi_{j,\hat{h}_2})(\hat{x}), & x \in \overline{\Omega_- \cup \Omega_+}, \\ 0, & \text{elsewhere,} \end{cases}$$

which will be associated to the grid points $h_- := F_-(\nu_{j,M_j}, \hat{h}_2)$ and $h_+ := F_+(\nu_{j,1}, \hat{h}_2)$, respectively, see the lower case Fig. 1.26 on p. 72.

Finally, we have to enforce the matching conditions along points $h \in \mathcal{H}_\sigma^{int}$, where $\hat{h}_2 \in \mathcal{J}_j$. In this case there is no scaling function associated to h. Again we are reduced to the univariate interface point case, but now in the situation considered in Subsect. 1.5.5. Considering the basis functions $\hat{\vartheta}_j^l$ defined in (1.188) for $l = -, 0, +$, we end up with three wavelets

$$\psi_{j,h_l}(x) = \begin{cases} (\hat{\vartheta}_j^l \otimes \eta_{j,\hat{h}_2})(\hat{x}), & x \in \overline{\Omega_- \cup \Omega_+}, \\ 0, & \text{elsewhere,} \end{cases}$$

which will be associated to the grid points $h_- := F_-(\nu_{j,M_j}, \hat{h}_2)$, $h_0 =: F_-(1, \hat{h}_2) = F_+(0, \hat{h}_2)$ and $h_+ := F_+(\nu_{j,1}, \hat{h}_2)$, see the upper case in Fig. 1.26 on p. 72.

1.5.7 Trivariate Matched Wavelets

In this section, the exposition will be deliberately less detailed than in the two previous sections, as the three dimensional construction follows the same spirit presented before. We shall be mainly concerned with the matching around a cross point. The matching around an edge or across a face will be easily reduced to lower dimensional situations.

Matched Wavelets Around a Cross Point. Let us first assume that the cross point C belongs to Ω. Let $N_C = N_{C,3}$ denote the number of subdomains meeting at C, and let these subdomains be (re-)labeled by $\Omega_1, \ldots, \Omega_{N_C}$. It is not restrictive to assume that for all $i \in \{1, \ldots, N_C\}$, one has $C = F_i(0,0,0)$.

It will be useful to express the number $N_{C,2}$ of faces and the number $N_{C,1}$ of edges meeting at C as a function of N_C. This can be accomplished as follows. Consider the tetrahedron in $\bar{\hat{\Omega}}$

$$\hat{T} := \{(\hat{x}_1, \hat{x}_2, \hat{x}_3) \ : \ \hat{x}_1 \geq 0,\, \hat{x}_2 \geq 0,\, \hat{x}_3 \geq 0,\, \hat{x}_1 + \hat{x}_2 + \hat{x}_3 \leq 1\}$$

and, for each $i \in \{1, \ldots, N_C\}$, set $T_i := F_i(\hat{T})$. Then, $P_C := \bigcup_i T_i$ is a (distorted) polyhedron in $\mathbb{R}^3$, with the following property: each face (or edge or vertex, resp.) of P_C is in one-to-one correspondence with a subdomain (or face or edge, resp.) meeting at C. Thus, by Euler's polyhedron Theorem, we get

$$N_{C,3} - N_{C,2} + N_{C,1} = 2.$$

On the other hand, since each face of P_C is a (distorted) triangle and each edge of P_C is shared by exactly two faces, one has $3N_{C,3} = 2N_{C,2}$. It follows that

$$N_{C,2} = \tfrac{3}{2} N_C, \qquad N_{C,1} = \tfrac{1}{2} N_C + 2. \tag{1.200}$$

The grid points $h \in \mathcal{H}_j$, to which we are going to associate the matched wavelets, have the form

$$h_{C,i,e} = F_i(\zeta_{e_1}, \zeta_{e_2}, \zeta_{e_3}), \qquad i \in \{1, \ldots, N_C\}, \quad e \in E^3 \setminus \{(0,0,0)\}, \tag{1.201}$$

with

$$\zeta_{e_l} := \begin{cases} 0, & \text{if } e_l = 0, \\ \nu_{j,1}, & \text{if } e_l = 1. \end{cases}$$

Note that there is exactly one of such points which lies inside each subdomain, each face and each edge meeting at C. Thus, the total number of such points is

$$N_{C,3} + N_{C,2} + N_{C,1} = 3N_C + 2.$$

This is precisely the number of wavelets to be constructed around C. Indeed, define for each $e \in E^3$

$$\psi_e^{(i)}(x) = \hat{\psi}_e(\hat{x}) = \theta_{e_1}(\hat{x}_1)\theta_{e_2}(\hat{x}_2)\theta_{e_3}(\hat{x}_3),$$

where

$$\theta_{e_l} := \begin{cases} \xi_{j,0}, & \text{if } e_l = 0, \\ \eta_{j,\nu_{j,1}}, & \text{if } e_l = 1. \end{cases}$$

Let us consider the spaces

$$\begin{aligned} S_{j+1}^C(\Omega_i) &:= \text{span}\{\psi_e^{(i)} \ : \ e \in E^3\} \\ &= \left\{ v^{(i)} = \sum_{e \in E^3} \alpha_e^{(i)} \psi_e^{(i)} \ : \ \boldsymbol{\alpha}^{(i)} := (\alpha_e^{(i)})_{e \in E^3} \in \mathbb{R}^3 \right\} \end{aligned}$$

and let us introduce the column vector $\boldsymbol{\alpha} := (\boldsymbol{\alpha}^{(i)})_{i=1,\dots,N_C} \in \mathbb{R}^{8N_c}$ (8 being the cardinality of E^3). In order to characterize the local space

$$S_{j+1}^C(\Omega) := \{v \in C^0(\bar{\Omega}) : v_{|\Omega_i} \in S_{j+1}^C(\Omega_i) \text{ if } i \in \{1, \dots, N_C\}, \\ v_{|\Omega_i} \equiv 0 \text{ elsewhere}\},$$

we enforce continuity firstly at C, secondly at the edges and finally at the faces meeting at C. This gives $N_C - 1$ conditions at C, which can be written as

$$\mathcal{C}_0\boldsymbol{\alpha} = \mathbf{0}, \qquad \text{where} \qquad \mathcal{C}_0 = \begin{bmatrix} \boldsymbol{c}_0 & -\boldsymbol{c}_0 & & \\ & \ddots & \ddots & \\ & & \boldsymbol{c}_0 & -\boldsymbol{c}_0 \end{bmatrix} \in \mathbb{R}^{(N_C-1)\times 8N_C},$$

with $\boldsymbol{c}_0 = (1,1,1,1,1,1,1,1)$. Next, we have $N_{ed} - 1$ conditions at each edge, where N_{ed} is the number of subdomains meeting at the edge; all these conditions can be written as $\mathcal{C}_1\boldsymbol{\alpha} = \mathbf{0}$, for a suitable matrix $\mathcal{C}_1$ whose structure depends on the topology of the subdomains. Finally, we have 1 condition at each face; they can be represented as $\mathcal{C}_2\boldsymbol{\alpha} = \mathbf{0}$.

In addition, we build functions in $W_j^C(\Omega) := \{v \in S_{j+1}^C(\Omega) : \langle v, \tilde{\varphi}_{j,C}\rangle_\Omega = 0\}$; this adds the condition $\mathcal{B}\boldsymbol{\alpha} = \mathbf{0}$, with $\mathcal{B} := (\boldsymbol{b}, \dots, \boldsymbol{b})$ and $\boldsymbol{b} = (1,0,0,0,0,0,0,0)$. Summarizing, we have

$$\mathcal{D}\,\boldsymbol{\alpha} = \mathbf{0} \qquad \text{with} \qquad \mathcal{D} = \begin{bmatrix} \mathcal{C}_0 \\ \mathcal{C}_1 \\ \mathcal{C}_2 \\ \mathcal{B} \end{bmatrix}.$$

All the conditions that we have enforced are linearly independent, as shown in [23] (see Sect. 5.2); thus, their number is

$$(N_C - 1) + \sum_{edges} (N_{ed} - 1) + N_{C,2} + 1 = N_C + 3N_C - N_{C,1} + N_{C,2} = 5N_C - 2.$$

Indeed, since each subdomain contains 3 edges meeting in C, one has

$$\sum_{edges} (N_{ed} - 1) = 3N_C - N_{C,1}.$$

We conclude that $\dim W_j^C(\Omega) = 3N_C + 2$, as desired. After the dual construction is made, biorthogonalization is accomplished as described in the previous sections; we omit the details. In conclusion, we end up with $3N_C + 2$ primal and dual wavelets $\psi_{j,C}^l$ and $\tilde{\psi}_{j,C}^l$, $l = 1, \dots, 3N_C + 2$; they are associated to the grid points (1.201) on p. 74, which from now on will be indicated by $h_{C,l}$.

Wavelets with Localized Support. Let us first notice that, as in lower dimension, no function in $W_j^C(\Omega)$ exists, which is supported in only one subdomain Ω_i. On the contrary, it is easily seen that for each couple of contiguous subdomains, two linearly independent functions in $W_j^C(\Omega)$ can be built, which vanish identically outside the two subdomains.

Thus, to each point $h_{C,l}$ which lies inside the common face of two subdomains, we associate one of such wavelet. To all but one points $h_{C,l}$ lying inside the subdomains, we associate other such wavelets, choosing them to be linearly independent from the previous ones. To each point $h_{C,l}$ lying inside an edge, we associate a wavelet supported in the closure of the union of all subdomains sharing the edge: it has the local structure of a tensor product of a two-dimensional scaling function $\varphi_{j,C}$ times the wavelet η_{j,ν_1} in the direction of the edge. Finally, to the remaining point $h_{C,l}$ interior to a subdomain, we associate the global wavelet.

Tensor Products of Matched Functions. According to (1.200), the number of subdomains meeting at C is even. We consider here the particular situation in which these subdomains can be grouped in two sets of equal cardinality $N_C/2$, all the subdomains of each set sharing a common edge stemming from C. For instance, this is the relevant case of 8 subdomains meeting at C, and representing the images of unit cubes lying in the 8 octants of $\mathbb{R}^3$.

By possibly introducing additional parametric mappings, we may reduce ourselves to the situation in which $N_C/2$ subdomains lie in the upper half space $\hat{x}_3 > 0$ while the remaining ones lie in the lower half plane. Each upper subdomain Ω_i can be written as $\Omega_i = \Omega_i' \times (0,1)$, where Ω_i' is a 2D subdomain in the plane $\hat{x}_3 = 0$; the companion lower subdomain is $\Omega_i' \times (-1,0)$. Thus, we are led to consider the case of $N_C/2$ subdomains in the plane meeting at $C' = (0,0)$. Let φ_j^{II} be the bivariate scaling function associated to C', and let $\psi_j^{II,l}$ $(l = 1,\dots,N_C/2)$ be any system of bivariate wavelets around C, built as in Subsect. 1.5.6 on p. 63. Moreover, let $\hat{\psi}_j^{I,l}$ $(l = -,+)$ be the univariate wavelets defined in Subsect. 1.5.5 on p. 58, and let $\hat{\vartheta}_j^{I,l}$ $(l = -,0,+)$ the univariate matched functions defined in Subsect. 1.5.5 on p. 61. Then, a system of wavelets around C can be defined as follows:

The functions $\psi_j^{II,l} \otimes \hat{\vartheta}_j^{I,+}$, with $l = 1,\dots,N_C/2$, are associated to the grid points having a strictly positive $\hat{x}_3$-component and not lying on the $\hat{x}_3$-axis; the functions $\psi_j^{II,l} \otimes \hat{\vartheta}_j^{I,0}$ and $\psi_j^{II,l} \otimes \hat{\vartheta}_j^{I,-}$ are associated to the analogous grid points having zero or negative $\hat{x}_3$-component. Finally, the functions $\varphi_j^{II} \otimes \hat{\psi}_j^{I,\pm}$ are associated to the remaining grid points on the $\hat{x}_3$-axis (obviously, these functions are extended by zero outside the union of the subdomains).

In the case of 8 subdomains meeting at C, the bivariate scaling and wavelet functions may be chosen to be themselves tensor products of univariate matched functions, so one can obtain a fully tensorized local wavelet basis around C.

Cross Points Lying on the Boundary. Let us now assume that the cross point C belongs to $\partial\Omega$. We follow the same notation as before. The grid points $h \in \mathcal{H}_j$ around C to which wavelets will be associated are again of the form (1.201) on p. 74, but now the points lying on a face or an edge contained in $\bar{\Gamma}_{\rm Dir}$ are missing. Let us denote by $N_{C,2}^{\rm Dir}$ ($N_{C,1}^{\rm Dir}$, resp.) the number of faces (edges, resp.) containing C and contained in $\bar{\Gamma}_{\rm Dir}$. It is easily seen that the number l_C of grid points $h_{C,l}$ we are interested in is

$$l_C = N_{C,3} + (N_{C,2} - N_{C,2}^{\rm Dir}) + (N_{C,1} - N_{C,1}^{\rm Dir}).$$

Let us count the number of conditions that define the space $W_j^C(\Omega)$. We have $N_{C,3} - 1$ matching conditions at C, plus one vanishing condition if $C \in \bar{\Gamma}_{\rm Dir}$ or one orthogonality condition if $C \notin \bar{\Gamma}_{\rm Dir}$. Next, we have $N_{ed} - 1$ matching conditions at each edge, plus one vanishing condition at each edge contained in $C \in \bar{\Gamma}_{\rm Dir}$. Finally, let N_f indicate the number of subdomains sharing the face f (this is 2 if the face is not contained in $\partial\Omega$, 1 if it is); then, we have $N_f - 1$ matching conditions at each face, plus one vanishing condition at each face contained in $C \in \bar{\Gamma}_{\rm Dir}$. Observing that

$$\sum_{edges} N_{ed} = \sum_{faces} N_f = 3N_{C,3},$$

the total number t_C of conditions which define $W_j^C(\Omega)$ is

$$t_C = 7N_{C,3} - (N_{C,2} - N_{C,2}^{\rm Dir}) - (N_{C,1} - N_{C,1}^{\rm Dir}).$$

Since these conditions are linearly independent (see again [23]), we obtain, as desired,

$$\dim W_j^C(\Omega) = 8N_{C,3} - t_C = l_C.$$

As in the interior cross point case, wavelets can be constructed with localized support. Actually, if $C \in \bar{\Gamma}_{\rm Dir}$, wavelets exist which are supported within one subdomain; they will be associated to the point $h_{C,l}$ lying inside the corresponding subdomain.

Matched Wavelets Around an Edge or a Face. Let $\sigma = ed$ be an edge, at which N_{ed} subdomains meet. We can reduce ourselves to the situation in which the subdomains are (re-)labeled by $\Omega_1, \dots, \Omega_{N_{ed}}$ and for each $i \in \{1, \dots, N_{ed}\}$ we have

$$ed = \{\, F_i(0,0,\zeta) \ : \ 0 \le \zeta \le 1 \,\}.$$

It is enough to consider points $h_{ed} \in \mathcal{K}_{j+1}$ which are internal to ed. Precisely, if $h_{ed} = F_i(0,0,\hat{h}_3)$ with $\hat{h}_3 \in I_j^{int}$, then we build wavelets which, in local coordinates, can be written as $\hat{\psi}_{j,ed}^l \otimes \xi_{j,\hat{h}_3}$, where $\hat{\psi}_{j,ed}^l$ are matched bivariate wavelets as defined in subsections 1.5.6 on p. 63 or 1.5.6 on p. 68. They will be associated to the points $h \in \mathcal{H}_j$ having the form $h = F_i(\zeta_1, \zeta_2, \hat{h}_3)$, with $\zeta_1, \zeta_2 \in \{0, \nu_{j,1}\}$.

On the other hand, if $h_{ed} = F_i(0, 0, \hat{h}_3)$ with $\hat{h}_3 \in J_j^{int}$, then the wavelets will be locally represented as $\psi^l_{j,ed} \otimes \eta_{j,\hat{h}_3}$, where now $\psi^l_{j,ed}$ are matched bivariate wavelets, defined as in the previously quoted subsections, but without enforcing the biorthogonality condition $\boldsymbol{\mathcal{B}\alpha} = \mathbf{0}$. The association to the grid points surrounding h_{ed} is done as above (note that now $h_{ed} \in \mathcal{H}_j$).

At last, let $\sigma = f$ be a face common to two subdomains Ω_- and Ω_+. The reference situation is such that

$$f = \{\, F_+(0, \zeta_2, \zeta_3) \ : \ 0 \leq \zeta_2, \zeta_3 \leq 1 \,\}.$$

Let $h_f \in \mathcal{K}_{j+1}$. If $h_f = F_+(0, \hat{h}_2, \hat{h}_3)$ with $\hat{h}_2, \hat{h}_3 \in I_j^{int}$, then we build wavelets having the local representation $\hat{\psi}^l_{j,f} \otimes \xi_{j,\hat{h}_2} \otimes \xi_{j,\hat{h}_3}$, where $\hat{\psi}^l_{j,f}$ are matched univariate wavelets as defined in Subsect. 1.5.5 on p. 58. On the other hand, if $h_f = F_+(0, \hat{h}_2, \hat{h}_3)$ with $\hat{h}_2, \hat{h}_3 \in I_j^{int} \cup J_j^{int}$ and at least one coordinate in J_j^{int}, then the local representation of the wavelets will be one of the following ones:

$$\hat{\vartheta}^l_{j,f} \otimes \eta_{j,\hat{h}_2} \otimes \xi_{j,\hat{h}_3}, \quad \hat{\vartheta}^l_{j,f} \otimes \xi_{j,\hat{h}_2} \otimes \eta_{j,\hat{h}_3}, \quad \hat{\vartheta}^l_{j,f} \otimes \eta_{j,\hat{h}_2} \otimes \eta_{j,\hat{h}_3},$$

where now $\hat{\vartheta}^l_{j,f}$ are matched univariate functions built in Subsect. 1.5.5. The association of these wavelets to the grid points surrounding h_f is straightforward.

1.5.8 Characterization of Sobolev Spaces

Vanishing Moments. Before progressing, let us make a general remark on vanishing moments. Let us first consider the case $\Gamma_{\mathrm{Dir}} = \emptyset$. Then, wavelets on the reference domain satisfy the conditions

$$\int_{\hat{\Omega}} \hat{x}^r \hat{\psi}_{j,\hat{h}}(\hat{x})\, d\hat{x} = 0, \qquad \hat{h}, \ \ |r| \leq \tilde{L} - 1, \tag{1.202}$$

where $\hat{x}^r = (\hat{x}_1^{r_1}, \ldots, \hat{x}_n^{r_n})$ and $|r| = \max_i r_i$; this follows from the fact that $\tilde{S}_j(\hat{\Omega})$ contains the set $\mathcal{P}_{\tilde{L}-1}(\hat{\Omega})$ of all polynomials of degree $\leq \tilde{L} - 1$ in each space variable. Unless a very special mapping is used, similar conditions in Ω are not satisfied. However, they are replaced by analogous conditions, involving a modified space. Indeed, $\tilde{S}_j(\Omega)$ contains the subspace

$$\mathcal{P}_{\tilde{L}-1}(\Omega) = \{p \in \mathcal{C}^0(\bar{\Omega}) \ : \ (p_{|\Omega_i})\hat{\ } \in \mathcal{P}_{\tilde{L}-1}(\hat{\Omega}), \ i = 1, \ldots, N\}, \tag{1.203}$$

so that one has

$$\langle p, \psi_{j,h} \rangle_\Omega = 0, \qquad h \in \mathcal{H}_j, \ \ p \in \mathcal{P}_{\tilde{L}-1}(\Omega). \tag{1.204}$$

A dual condition holds for the dual wavelets $\tilde{\psi}_{j,h}$. These conditions still imply the compression properties of wavelets, to be understood in the following sense: the wavelet coefficients $f_{j,k} := \langle f, \psi_{j,k}\rangle_\Omega$ of a function in $L_2(\Omega)$ decay rapidly if f is smooth.

Let us now assume that $\Gamma_{\mathrm{Dir}} \neq \emptyset$. Then, since our dual scaling functions vanish on Γ_{Dir}, orthogonality to the constants is lost for *some* of our primal wavelets $\psi_{j,h}$. Precisely, this happens for the wavelets whose representation on the reference domain is $\hat{\psi}_{j,\hat{h}} = \bigotimes_{i=1}^n \hat{\vartheta}_{\hat{h}_i}$ and there exists a nonempty set $I \subseteq \{1,\ldots,n\}$ such that $\hat{\vartheta}_{\hat{h}_i} = \hat{\eta}^D_{\hat{h}_i}$ for $i \in I$ and $\hat{\vartheta}_{\hat{h}_i} = \hat{\xi}_{j,\hat{h}_i}$ for $i \notin I$. Note that the number of such wavelets is asymptotically negligible as compared to the cardinality of the full wavelet basis Ψ_j on Ω.

However, if orthogonality to constants is needed, dual scaling functions which do not vanish on Γ_{Dir} have to be built. Such a construction has recently been accomplished in [58] on the reference cube, and it can easily be incorporated in our domain decomposition method, by slightly adapting the matching around Dirichlet and mixed boundary cross points described below.

It can be shown that all wavelets supported in two subdomains also have patchwise vanishing moments. However, this is not true for functions around cross points that are supported in all adjacent subdomains.

Norm Equivalences. At the end of our construction we shall obtain a system of biorthogonal wavelets on Ω which allows the characterization of certain smoothness spaces. For instance, let us set

$$H^s_b(\Omega) := \{v \in H^s(\Omega) \,:\, v = 0 \text{ on } \Gamma_{\mathrm{Dir}}\}, \qquad s \in \mathbb{N} \setminus \{0\}, \tag{1.205}$$

and let us extend the definition by interpolation for $s \notin \mathbb{N}$, $s > 0$ (after setting $H^0_b(\Omega) = L_2(\Omega)$). Furthermore, we introduce another scale of Sobolev spaces, depending upon the partition $\mathcal{P} := \{\Omega_i \,:\, i = 1,\ldots,N\}$ of Ω; precisely, we set

$$H^s_b(\Omega;\mathcal{P}) := \{v \in H^1_b(\Omega) \,:\, v_{|\Omega_i} \in H^s(\Omega_i), i = 1,\ldots,N\} \tag{1.206}$$

for $s \in \mathbb{N} \setminus \{0\}$, equipped with the norm

$$\|v\|_{H^s_b(\Omega;\mathcal{P})} := \left(\sum_{i=1}^N \|v_{|\Omega_i}\|^2_{H^s(\Omega_i)}\right)^{1/2}, \qquad v \in H^s_b(\Omega;\mathcal{P}),$$

and we extend the definition using interpolation for $s \notin \mathbb{N}$, $s > 0$ (again we set $H^0_b(\Omega;\mathcal{P}) = L_2(\Omega)$). The following theorem summarizes the characterization features of our wavelet systems; they can be exploited in many different applications.

Theorem 5. ([23], Theorem 5.6) *Assume that* $s \in (-\tilde{\gamma}, \gamma)$, *where* $\tilde{\gamma} = \tilde{t}$ *in* (1.13) *on p. 6 and* $\gamma = d - 1/2$ *in* (1.11) *on p. 5. Then*

$$H^s_b(\Omega;\mathcal{P}) = \left\{ v \in L_2(\Omega) \ : \ \sum_{j=j_0}^{\infty} \sum_{h\in\mathcal{H}_j} 2^{2sj}\, |\langle v, \tilde{\psi}_{j,h}\rangle_\Omega|^2 < \infty \right\}. \quad (1.207)$$

In addition, if $v \in H^s_b(\Omega;\mathcal{P})$, then

$$v = \sum_{k\in\mathcal{K}_{j_0}} \langle v, \tilde{\varphi}_{j_0,k}\rangle_\Omega\, \varphi_{j_0,k} \ + \sum_{j=j_0}^{\infty} \sum_{h\in\mathcal{H}_j} \langle v, \tilde{\psi}_{j,h}\rangle_\Omega\, \psi_{j,h}, \quad (1.208)$$

the series being convergent in the norm of $H^s_b(\Omega;\mathcal{P})$, and

$$\|v\|^2_{H^s_b(\Omega;\mathcal{P})} \sim \sum_{k\in\mathcal{K}_{j_0}} 2^{2sj_0}\, |\langle v, \tilde{\varphi}_{j_0,k}\rangle_\Omega|^2 \ + \sum_{j=j_0}^{\infty} \sum_{h\in\mathcal{H}_j} 2^{2sj}\, |\langle v, \tilde{\psi}_{j,h}\rangle_\Omega|^2 \quad (1.209)$$

for the above range of s. A dual statement holds if we exchange the roles of $S_j(\Omega)$ and $\tilde{S}_j(\Omega)$. □

Finally, let us comment on the range of applicability of the constructed wavelet bases. For $-1/2 < s < 3/2$, one has $\|\cdot\|_{H^s_b(\Omega;\mathcal{P})} \sim \|\cdot\|_{s,\Omega}$. In fact, the particular choice of $\|\cdot\|_{H^s_b(\Omega;\mathcal{P})}$ is induced by the construction principle 'mapping and matching'. As we shall see later, this allows to precondition an elliptic operator acting on such a space $H^s(\Omega)$ in the sense that the operator equation can equivalently be written as a boundedly invertible ℓ_2-problem. However, the above equivalence of norms does not hold for $s \le 1/2$ and $s \ge 3/2$. This restricts the scope of problems to $-1/2 < s < 3/2$, see also the remarks at the beginning of Sect. 1.5 on p. 47 and Sect. 3.1.1 on p. 109 below.

Refinement Matrices. For some applications it is advantageous to have explicit access to the representation of the wavelet bases Ψ_j, $\tilde{\Psi}_j$ in terms of the scaling functions system, i.e.,

$$\Psi_j = \boldsymbol{M}^T_{j,1}\, \Phi_{j+1}, \qquad \tilde{\Psi}_j = \tilde{\boldsymbol{M}}^T_{j,1}\, \tilde{\Phi}_{j+1},$$

see also (1.22) on p. 8. The matrices $\boldsymbol{M}_{j,1}$, $\tilde{\boldsymbol{M}}_{j,1}$ are often called *refinement matrices* of the wavelet bases. Even though we do not determine these matrices explicitly, they can easily be deduced in the following way. The refinement matrices for the univariate bases are known explicitly, see Sect. 1.3.4 above. Moreover, we derive *explicit* formulas for the matching coefficients, i.e., the representation of the wavelets in Ω in terms of those in the subdomains, which are images of tensor product functions. Hence, by putting the pieces together, this immediately gives the refinement coefficients for the wavelets. This procedure has been used for numerically solving the Helmholtz problem

$$u \in H^1_0(\Omega): \quad (\nabla u, \nabla v)_{0,\Omega} + c(u,v)_{0,\Omega} = (f,v)_{0,\Omega} \quad \text{for all } v \in H^1_0(\Omega)$$

on the L-shaped domain by the Wavelet Element Method, see [22].

1.6 Vector Wavelets

For the space $\boldsymbol{L}_2(\Omega)$ of square integrable vector fields, we denote wavelet systems by boldface characters, i.e., $\boldsymbol{\Psi}$, and in (1.4) one uses the standard norm

$$\|\boldsymbol{f}\|_{0,\Omega}^2 := \sum_{i=1}^{n} \|f_i\|_{0,\Omega}^2$$

for $\boldsymbol{f} \in \boldsymbol{L}_2(\Omega)^n$. Moreover, in many cases, we have to equip the index $\lambda \in \mathcal{J}$ labeling the scalar wavelets with some additional index indicating the component of the vector field. For example, let $\Psi^{[\nu]} := \{\psi_\lambda^{[\nu]} : \lambda \in \mathcal{J}^{[\nu]}\}$, $\tilde{\Psi}^{[\nu]} := \{\tilde{\psi}_\lambda^{[\nu]} : \lambda \in \mathcal{J}^{[\nu]}\}$, $1 \le \nu \le n$, be (possibly different) biorthogonal systems in $L_2(\Omega)$. Then, the vector fields

$$\boldsymbol{\psi}_{(i,\lambda)} := \psi_\lambda^{[i]}\, \boldsymbol{\delta}_i, \qquad \tilde{\boldsymbol{\psi}}_{(i,\lambda)} := \tilde{\psi}_\lambda^{[i]}\, \boldsymbol{\delta}_i, \qquad \lambda \in \mathcal{J}^{[i]},\ 1 \le i \le n,$$

obviously form a biorthogonal wavelet basis for $\boldsymbol{L}_2(\Omega)$. Here we denote by $\boldsymbol{\delta}_i := (\delta_{1,i}, \ldots, \delta_{n,i})^T$, $1 \le i \le n$ the canonical unit vector in $\mathbb{R}^n$. Denoting by

$$\boldsymbol{\mathcal{J}} := \bigcup_{i=1}^{n} \bigcup_{\lambda \in \mathcal{J}^{[i]}} (i, \lambda), \qquad \boldsymbol{\lambda} := (i, \lambda),$$

the corresponding set of indices, we obtain the following biorthogonal vector wavelet system

$$\boldsymbol{\Psi} = \{\boldsymbol{\psi}_{\boldsymbol{\lambda}} : \boldsymbol{\lambda} \in \boldsymbol{\mathcal{J}}\}, \qquad \tilde{\boldsymbol{\Psi}} = \{\tilde{\boldsymbol{\psi}}_{\boldsymbol{\lambda}} : \boldsymbol{\lambda} \in \boldsymbol{\mathcal{J}}\}.$$

2 Wavelet Bases for $\boldsymbol{H}(\text{div})$ and $\boldsymbol{H}(\text{curl})$

In this chapter, we follow mostly [118] and construct wavelet bases specifically for the spaces $\boldsymbol{H}(\text{div};\Omega)$ and $\boldsymbol{H}(\textbf{curl};\Omega)$. These spaces arise naturally in the variational formulation of a whole variety of partial differential equations. Two prominent examples are the *Navier-Stokes equations* that describe the flow of a viscous, incompressible fluid and *Maxwell's equations* in electromagnetism. For the incompressible Navier-Stokes equations, $\boldsymbol{H}(\text{div};\Omega)$ plays an important role for modeling the velocity-field of the flow. The space $\boldsymbol{H}(\textbf{curl};\Omega)$ has to be considered, when one is interested in a formulation in non-primitive variables such as stream function, vorticity and vector potential, [77]. Certain electromagnetic phenomena are known to be modeled by Maxwell's equations. Here, the space $\boldsymbol{H}(\textbf{curl};\Omega)$ appears when linking the quantities electric and magnetic field, magnetic induction and flux density, see for example [10, 17, 86], the references therein and also Chap. 3 on p. 109 below.

Here, we construct a biorthogonal wavelet basis for $\boldsymbol{H}(\textbf{curl};\Omega)$ and introduce **curl**-free vector wavelets. It is well known, that the spaces $\boldsymbol{V}(\text{div};\Omega)$ and $\boldsymbol{V}(\textbf{curl};\Omega)$ of divergence- resp. **curl**-free vector fields in $\boldsymbol{L}_2(\Omega)$ are linked by certain interrelations. These play an important role when one considers the splitting of $\boldsymbol{L}_2(\Omega)$ into $\boldsymbol{V}(\text{div};\Omega)$, $\boldsymbol{V}(\textbf{curl};\Omega)$, respectively, and some complement. Using divergence-free wavelets [98, 116] it was shown that there exists a stable (biorthogonal) Helmholtz decomposition of $\boldsymbol{H}(\text{div};\Omega)$, [117], which, however is not an orthogonal decomposition. Now, with the construction of wavelet bases in $\boldsymbol{H}(\textbf{curl};\Omega)$ at hand, we can give an explicit *orthogonal* decomposition of $\boldsymbol{L}_2(\Omega)$.

The key ingredient for the construction described in this chapter is a certain relation of different wavelet systems through differentiation and integration. We are going to describe this relation in Sect. 2.1. After collecting the basic definitions and notation in Sect. 2.2 on p. 90, Sect. 2.3 on p. 92 and 2.4 on p. 98 contain the actual construction. Sect. 2.5 is devoted to the derivation of Hodge decompositions. We will perform the construction under general assumptions on the used wavelet systems on the domain Ω under consideration (see Sect. 2.1.3 on p. 86). It turns out that these assumptions mainly pose conditions on the geometry of Ω. Hence, we will investigate how our construction can be performed on general domains in Sect. 2.6 on p. 102. Finally, in Sect. 2.7 on p. 105, we describe a particular example.

2.1 Differentiation and Integration

We are now going to describe a fundamental relation between different biorthogonal systems. In fact, as we shall see certain biorthogonal wavelets are linked by differentiation and integration. We will first review the results for wavelet systems on the real line and on $(0,1)$ and then we formulate a general assumption that can be interpreted as a generalization of the known results.

2.1.1 Differentiation and Integration on the Real Line

The following fundamental relations of certain wavelet systems on the real line, differentiation and integration has been proven by P.G. Lemarié-Rieusset [98]: Pics/wav11.eps

Theorem 6. *Let $\eta^{(1)}$, $\tilde{\eta}^{(1)} \in L_2(\mathbb{R})$ be biorthogonal wavelets on $\mathbb{R}$ induced by dual scaling functions $\xi^{(1)}$, $\tilde{\xi}^{(1)}$, such that $\xi^{(1)}$, $\eta^{(1)} \in H_0^1(\mathbb{R})$. Then, there exists a second pair $\xi^{(0)}$, $\tilde{\xi}^{(0)}$ of dual scaling functions and associated biorthogonal wavelets $\eta^{(0)}$, $\tilde{\eta}^{(0)}$ fulfilling*

$$\frac{d}{dx}\xi^{(1)}(x) = \xi^{(0)}(x) - \xi^{(0)}(x-1), \qquad \frac{d}{dx}\tilde{\xi}^{(0)}(x) = \tilde{\xi}^{(1)}(x+1) - \tilde{\xi}^{(1)}(x),$$
$$\frac{d}{dx}\eta^{(1)}(x) = 4\,\eta^{(0)}(x), \qquad \frac{d}{dx}\tilde{\eta}^{(0)}(x) = (-4)\,\tilde{\eta}^{(1)}(x). \qquad \square \tag{2.1}$$

This means that starting by a pair $\xi^{(1)}$, $\tilde{\xi}^{(1)}$ where $\xi^{(1)}$ is sufficiently smooth, we obtain a new pair $\xi^{(0)}$, $\tilde{\xi}^{(0)}$ by the above procedure. For biorthogonal B-spline wavelets, these functions are explicitly known, namely if

$$\xi^{(1)} := {}_d\xi, \qquad \tilde{\xi}^{(1)} := {}_{d,\tilde{d}}\tilde{\xi},$$

we obtain

$$\xi^{(0)} := {}_{d-1}\xi, \qquad \tilde{\xi}^{(0)} := {}_{d-1,\tilde{d}+1}\tilde{\xi},$$

i.e., one looses one degree of smoothness but wins one additional moment for the primal wavelets. Due to different symmetry properties of even and odd indexed B-splines, the relations in (2.1) take the special form

$$\begin{aligned}\frac{d}{dx}{}_d\xi(x) &= {}_{d-1}\xi(x+\mu(d-1)) - {}_{d-1}\xi(x-\mu(d-1)),\\ \frac{d}{dx}{}_{d-1,\tilde{d}+1}\tilde{\xi}(x) &= {}_{d,\tilde{d}}\tilde{\xi}(x+1-\mu(d-1)) - {}_{d,\tilde{d}}\tilde{\xi}(x-\mu(d-1)),\end{aligned} \tag{2.2}$$

where again $\mu(d) := d \bmod 2$ (see (1.39) on p. 14). Let us refer again to Fig. 1.3 on p. 16. Considering the third and fourth columns, we can start the above process with the parameters $d = \tilde{d} = 3$ corresponding to the third row. As we can see, the smoothness of the primal wavelet decreases when going from the third row to the second one which corresponds to $\eta^{(0)}$ in this case, i.e., $d = 2$ and $\tilde{d} = 4$. On the contrary, the smoothness of the dual wavelet is increased. We can apply the process again for $d = 2$ and $\tilde{d} = 4$ as the starting functions and obtain the functions displayed in the first row corresponding to $d = 1$ and $\tilde{d} = 5$.

Of course, the process described in Theorem 6 can also be applied for orthonormal wavelets as long as they are sufficiently smooth. This means that $\xi^{(1)} = \tilde{\xi}^{(1)}$. In view of the relations (2.1) however it is obvious that $\xi^{(0)}$ and $\tilde{\xi}^{(0)}$ are different, but still biorthogonal. Hence one obtains a biorthogonal system even when starting with an orthonormal one.

To illustrate this, we have shown the resulting functions for $\xi^{(1)} = \tilde{\xi}^{(1)} = {}_3\xi$ in Fig. 2.1. The filters of the new systems can be obtained by the relations

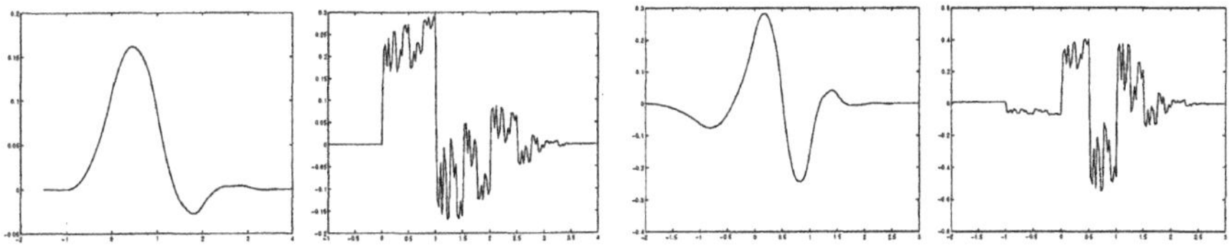

Fig. 2.1. Arising functions $\xi^{(0)}$, $\tilde{\xi}^{(0)}$, and $\eta^{(0)}$, $\tilde{\eta}^{(0)}$ for the orthonormal scaling functions and wavelets $\xi^{(1)} = {}_3\xi$, $\eta^{(1)} = {}_3\eta$, see Sect. 1.2.

$$\begin{aligned} a^{(0)}(z) &= a^{(1)}(z)\frac{2}{1+z}, & \tilde{a}^{(0)}(z) &= \tilde{a}^{(1)}(z)\frac{1+\bar{z}}{2}, \\ b^{(0)}(z) &= b^{(1)}(z)\frac{1-z}{2}, & \tilde{b}^{(0)}(z) &= \tilde{b}^{(1)}(z)\frac{2}{1-\bar{z}}, \end{aligned} \tag{2.3}$$

where

$$a(z) := \sum_k a_k\, z^k, \quad b(z) := \sum_k b_k\, z^k, \quad z \in \mathbb{C},$$

are the *symbols* of the masks $\{a_k\}_k$ and $\{b_k\}_k$, respectively.

2.1.2 Differentiation and Integration on $(0, 1)$

Now we review from [118] the generalization of Theorem 6 to wavelet systems on $(0, 1)$.

Theorem 7. *Let $\Xi_j^{(1)}$, $\tilde{\Xi}_j^{(1)}$ be a biorthogonal system of univariate B-spline scaling functions and $\Upsilon_j^{(1)}$, $\tilde{\Upsilon}_j^{(1)}$ the induced biorthogonal wavelet system on*

$(0,1)$ *such that* $\Upsilon^{(1)} \subset H_0^1((0,1))$. *Then, there exists a second system of dual scaling functions* $\Xi_j^{(0)}$, $\tilde{\Xi}_j^{(0)}$ *and induced biorthogonal wavelets* $\Upsilon_j^{(0)}$, $\tilde{\Upsilon}_j^{(0)}$ *(w.r.t. the same set of indices* I_j, J_j, *respectively) such that*

$$
\begin{aligned}
\tfrac{d}{dx}\Xi_j^{(1)} &= D_{j,0}\,\Xi_j^{(0)}, & \tfrac{d}{dx}\tilde{\Xi}_j^{(0)} &= -D_{j,0}^T\,\tilde{\Xi}_j^{(1)}, \\
\tfrac{d}{dx}\Upsilon_j^{(1)} &= D_{j,1}\,\Upsilon_j^{(0)}, & \tfrac{d}{dx}\tilde{\Upsilon}_j^{(0)} &= -D_{j,1}^T\,\tilde{\Upsilon}_j^{(1)},
\end{aligned}
\tag{2.4}
$$

where $D_{j,e} \in \mathbb{R}^{|J_{j,e}| \times |J_{j,e}|}$, $e = 0, 1$, *are sparse matrices such that* $D_{j,1}$ *is regular and* $\|D_{j,1}^{-1} D_{j,0}\| \lesssim 1$ *independent of* j. $\square$

Let us add some remarks on the structure of the matrices $D_{j,i}$ and the support of the wavelets. For interior functions, we obtain by Theorem 6 that

$$
(D_{j,0})_{k,k'} = 2^j(\delta_{k,k'} - \delta_{k,k'-1}), \qquad (D_{j,1})_{k,k'} = 2^{j+2}\delta_{k,k'}. \tag{2.5}
$$

According to the modifications near the boundaries we obtain block-diagonal matrices of the form

$$
D_{j,i} = \boxed{\begin{array}{ccc} D^L & & \\ & D^I & \\ & & D^R \end{array}} \in \mathbb{R}^{|J_{j,i}| \times |J_{j,i}|}, \; i = 0, 1. \tag{2.6}
$$

The large central block D^I is of the form (2.5) and the two blocks D^L, D^R that correspond to the modifications of those functions living near the boundary, are small. For $D_{j,1}$, it is obvious that the inverse is (block-) diagonal.

In Fig. 2.2, the sparsity pattern of the matrices $D_{j,0}$, $D_{j,1}$ and $D_{j,1}^{-1}$ are displayed. Some care is needed if the boundary conditions have to be incorporated into the systems. Let us illustrate this by the following example. Suppose that $\mathcal{P}_1 \subset S(\Xi_j^{(1)})$, then, in view of (2.4), the constant functions have to be included in $S(\Xi_j^{(0)})$. Hence, in this case, one cannot enforce homogeneous Dirichlet boundary conditions to $\Xi_j^{(0)}$. However, the above mentioned *complementary boundary* conditions [58] turn out to be the right choice (see Sect. 1.3.7 on p. 44).

2.1.3 Assumptions for General Domains

For the construction of wavelet bases for $H(\mathrm{div}; \Omega)$ and $H(\mathrm{curl}; \Omega)$, we need a suitable generalization of the results in Sect. 2.1.1 and 2.1.2 to the n-dimensional case, partial derivatives and domains $\Omega \subset \mathbb{R}^n$. Of course, we cannot expect that such a property holds for all possible wavelet constructions and all possible domains Ω, it has to be checked for each particular case. Hence, we formulate this generalization as an assumption.

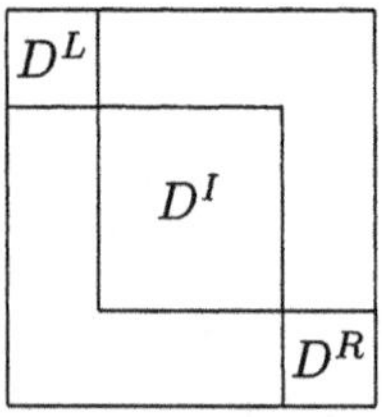

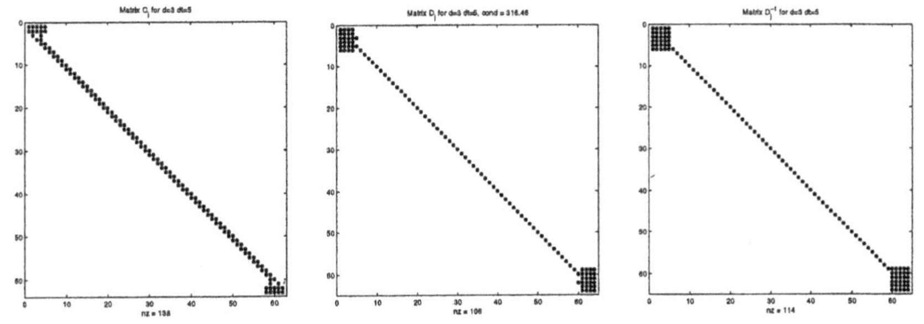

Fig. 2.2. Matrices $C_j = D_{j,0}$, $D_j = D_{j,1}$ and D_j^{-1} for $d =$ and $\tilde{d} = 5$.

Assumption 3. *For all $\gamma = (\gamma_1, \dots, \gamma_n)^T \in E^n := \{0,1\}^n$, and all $i \in \{1, \dots, n\}$ such that $\gamma - \delta_i \in E^n$, there exist biorthogonal systems $\Psi^{(\gamma)}$, $\tilde{\Psi}^{(\gamma)}$ on $L_2(\Omega)$ with the following properties:*

(a) *$\Psi^{(\gamma)}$, $\tilde{\Psi}^{(\gamma)}$ correspond to the* same *set of indices $\mathcal{J} = \{(j,k) : j \geq j_0,\ k \in \mathcal{J}_j\}$ for all $\gamma \in E^n$, where $j_0 \in \mathbb{Z}$ denotes some coarse level.*

(b) *The functions $\psi_\lambda^{(\gamma)}$, $\lambda \in \mathcal{J}$ are compactly supported.*

(c) *One has*

$$\partial_i \Psi^{(\gamma)} = D^{(i)} \Psi^{(\gamma - \delta_i)}, \tag{2.7}$$

where here $\Psi^{(\gamma)}$ is viewed as a column vector of functions and $D^{(i)}$ are sparse *transformations. By 'sparse' we mean here that for each $\lambda \in \mathcal{J}$, there is only a small and fixed number of non-zero entries $D_{\lambda,\mu}^{(i)}$ of $D^{(i)}$ ($\mu \in \mathcal{J}$). Here and in the sequel, we use the short hand notation $\partial_i := \frac{\partial}{\partial x_i}$ for $1 \leq i \leq n$ and we assume the partial derivatives in (2.7) to exist in the weak sense in $L_2(\Omega)$.*

(d) *For all $\lambda \in \mathcal{J}$ with $|\lambda| > j_0$ (i.e., ψ_λ is* not *a coarse level function), there exists an index $i_\lambda \in \{1, \dots, n\}$ such that*

$$D_{i_\lambda}^{(i_\lambda)} := (D_{\mu,\mu'}^{(i_\lambda)})_{\mu,\mu' \in \mathcal{J}_{|\lambda|};\, i_\mu = i_{\mu'} = i_\lambda} \tag{2.8}$$

is invertible and

$$\|(D_{i_\lambda}^{(i_\lambda)})^{-1}\, D_{i_\mu}^{(i_\mu)}\| \lesssim 1$$

holds for all for all $\mu \in \mathcal{J}_{|\lambda|}$ independent of $|\lambda|$, where $\|\cdot\|$ denotes some matrix norm.

(e) *We assume that $D_{\lambda,\mu}^{(i)} \neq 0$ only if $|\mu| = |\lambda|$ and $i_\lambda = i_\mu$.*

Let us add some comments on Ass. 3, in particular on the index i_λ. To clarify the idea, let us assume that Ω is a tensor product domain and that Ψ is a tensor product basis. If $\lambda \in \mathcal{J}$ corresponds to a wavelet function (i.e., $|\lambda| > j_0$), then at least one component of $\psi_\lambda^{(\gamma - \delta_{i_\lambda})}$ is a 1D wavelet (and *not* a scaling function). Hence, we can make use of the fact that $D_{|\lambda|,1}$ in Theorem 7 is invertible in order to express $\psi_\lambda^{(\gamma-\delta_{i_\lambda})}$ in terms of $\partial_{i_\lambda}\Psi_{|\lambda|}^{(\gamma)}$, which boils down to express $\eta_{\lambda_{i_\lambda}}^{(0)}$ in terms of $\frac{d}{dx}\Upsilon_{|\lambda|}^{(1)}$. This is precisely what the assumption in (2.8) states.

The restriction $i_\mu = i_{\mu'} = i_\lambda$ in (2.8) means that in the representation of a $\partial_{i_\lambda}\psi_\lambda^{(\gamma)}$ only those functions $\psi_\mu^{(\gamma-\delta_{i_\lambda})}$ appear that have the same index i_λ. This is a very reasonable assumption in view of Theorem (7) since it was shown there that $D_{j,1}$ is invertible.

Moreover, let us illustrate this assumption by Fig. 2.3 below for $n = 2$. Ass. 3 states here —roughly speaking— that we can apply partial derivatives in the indicated way. Moreover, it states that the partial derivative for each wavelet can be expressed by a finite linear combination of few wavelets of a 'lower' system. The conditions (d) and (e) are of more technical nature. However, they are very important for the subsequent construction, since they allow to express each wavelet in terms of linear combinations of certain partial derivatives of a 'higher' system. The easiest example to think of occurs for $\Omega = \mathbb{R}^2$ and simply taking tensor products of the functions arising in Theorem 6.

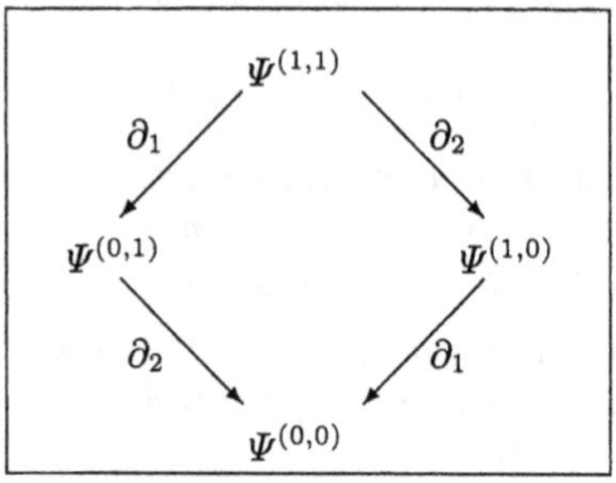

Fig. 2.3. Relationships between wavelet systems for $n = 2$ according to Ass. 3.

Note that condition (a) is by no means automatically fulfilled. It is a matter of adjusting the parameters that enter the construction of wavelet bases on bounded domains Ω (e.g. the parameter ℓ in Lemma 1 on p. 19). However, for a significantly large set of examples this problem has been solved in [33, 58, 117], see Sect. 2.1.2. We will also focus on this later in Sect. 2.6 below, where we give concrete examples.

Ass. 3 has some important immediate consequences, which will be used in the sequel: For each $\lambda \in \mathcal{J}$, we obtain by (2.7)

$$\partial_i \psi_\lambda^{(\gamma)} = \sum_{\mu \in \mathcal{J}_{|\lambda|}} D_{\lambda,\mu}^{(i)} \psi_\mu^{(\gamma-\delta_i)} = [D_{|\lambda|}^{(i)}]^\lambda \Psi^{(\gamma-\delta_i)}, \tag{2.9}$$

as well as $\partial_{i_\lambda}\psi_\lambda^{(\gamma)} = [D_{i_\lambda}^{(i_\lambda)}]^\lambda \Psi^{(\gamma-\delta_{i_\lambda})}$, where $[A]^\lambda$ denotes the λ-th row of A. Moreover, for the biorthogonal system $\tilde{\Psi}^{(\gamma)}$, we have

$$\partial_i \tilde{\Psi}^{(\gamma-\delta_i)} = -(D^{(i)})^T \tilde{\Psi}^{(\gamma)}, \tag{2.10}$$

which is easily seen. In fact, using the notation $\langle \Theta, \Phi \rangle := \big((\theta,\phi)_{L_2(\Omega)}\big)_{\theta\in\Theta,\phi\in\Phi}$ for two countable systems of functions, we obtain by integrating by parts and taking the compact support of $\psi_\lambda^{(\gamma)}$ into account:

$$D^{(i)} = D^{(i)}\langle \Psi^{(\gamma-\delta_i)}, \tilde{\Psi}^{(\gamma-\delta_i)}\rangle = \langle \partial_i \Psi^{(\gamma)}, \tilde{\Psi}^{(\gamma-\delta_i)}\rangle = (-1)\langle \Psi^{(\gamma)}, \partial_i \tilde{\Psi}^{(\gamma-\delta_i)}\rangle.$$

Now, due to (2.7) and the biorthogonality, we know that $\partial_i \tilde{\Psi}^{(\gamma-\delta_i)}$ is some linear combination of $\tilde{\Psi}^{(\gamma)}$, which shows (2.10). Hence, the corresponding biorthogonal projectors fulfill

$$\partial_i Q^{(\gamma)} v = Q^{(\gamma-\delta_i)}(\partial_i v), \qquad \partial_i Q_j^{(\gamma)} v = Q_j^{(\gamma-\delta_i)}(\partial_i v),\ j \geq 0. \tag{2.11}$$

2.1.4 Norm Equivalences

Under Ass. 3 we can now derive a different way to estimate the $H^1(\Omega)$-norm of a function than the 'standard' one given in (1.26) on p. 9 for $s = 1$. It will turn out later that this new estimate is in certain cases better suited for characterizing the spaces $\boldsymbol{H}(\mathrm{div};\Omega)$ and $\boldsymbol{H}(\mathbf{curl};\Omega)$. Moreover, in Chap. 3 on p. 109, this is used to derive robust and asymptotically optimal preconditioners for differential operators in these spaces involving physical constants.

Denoting by $\ell(\mathcal{J})$ the space of all series that are labeled by the set of indices $\mathcal{J}$, we define the operator $\mathsf{d}^{(i)} : \ell(\mathcal{J}) \to \ell(\mathcal{J})$, $i = 1,\ldots,n$, by

$$(\mathsf{d}^{(i)}\boldsymbol{c})_\lambda := \sum_{\mu\in\mathcal{J}_{|\lambda|}} D_{\mu,\lambda}^{(i)} c_\mu = [D_{|\lambda|}^{(i)}]_\lambda\, \boldsymbol{c}, \qquad \boldsymbol{c}\in\ell(\mathcal{J}), \tag{2.12}$$

where $[A]_\lambda$ denotes the λ-th column of A. Moreover, $\mathsf{d}^{(i_\lambda)}$ is defined in a straightforward manner. Then, we obtain

Remark 4. Under the hypotheses of Ass. 3, we have for each function

$$f = \sum_{\lambda\in\mathcal{J}} c_\lambda \psi_\lambda^{(\mathbf{1})} \in H^1(\Omega),$$

$\mathbf{1} := (1,\ldots,1)^T \in \mathbb{N}^n$, the estimate

$$\|f\|_{1,\Omega} \sim \|\boldsymbol{c}\|_{h^1(\mathcal{J})} := (\|\boldsymbol{c}\|_{\ell_2(\mathcal{J})}^2 + \|\mathsf{grad}\,\boldsymbol{c}\|_{\ell_2(\mathcal{J})}^2)^{1/2}, \tag{2.13}$$

where

$$\mathsf{grad}\,\boldsymbol{c} := (\mathsf{d}^{(1)}\boldsymbol{c},\ldots,\mathsf{d}^{(n)}\boldsymbol{c})^T \in \ell(\mathcal{J}) \tag{2.14}$$

and hence

$$\|\mathsf{grad}\,\mathbf{c}\|^2_{\ell_2(\mathcal{J})} = \sum_{i=1}^{n} \|\mathsf{d}^{(i)}\,\mathbf{c}\|^2_{\ell_2(\mathcal{J})}.$$

In fact,

$$\begin{aligned} \partial_i f &= \sum_{\lambda\in\mathcal{J}} \sum_{\mu\in\mathcal{J}_{|\lambda|}} c_\lambda\, D^{(i)}_{\lambda,\mu}\, \psi^{(1-\delta_i)}_\mu = \sum_{\lambda\in\mathcal{J}} \Big(\sum_{\mu\in\mathcal{J}_{|\lambda|}} D^{(i)}_{\mu,\lambda}\, c_\mu \Big) \psi^{(1-\delta_i)}_\lambda \\ &= \sum_{\lambda\in\mathcal{J}} (\mathsf{d}^{(i)}\,\mathbf{c})_\lambda\, \psi^{(1-\delta_i)}_\lambda . \end{aligned} \tag{2.15}$$

Note that we may change the ordering of the sums in the second equality above due to Ass. 3, (c) and (e). Using the norm equivalence (1.4) on p. 2 for $\Psi^{(1)}$ and $\Psi^{(1-\delta_i)}$, $1 \le i \le n$, yields

$$\begin{aligned} \|f\|^2_{1,\Omega} &= \|f\|^2_{0,\Omega,} + \|\mathbf{grad}\, f\|^2_{0,\Omega} \\ &\sim \|\mathbf{c}\|^2_{\ell_2(\mathcal{J})} + \sum_{i=1}^{n} \|\mathsf{d}^{(i)}\,\mathbf{c}\|^2_{\ell_2(\mathcal{J})} = \|\mathbf{c}\|^2_{h^1(\mathcal{J})}, \end{aligned}$$

which proves (2.13). □

We compare the estimates (2.13) and (1.26) on p. 9 by the following

Example 2. Let us consider the univariate case $\Omega = (0,1)$ and choose a wavelet $\psi^{(1)}_\lambda$, $\lambda = (j,k)$, such that

$$\psi^{(1)}_\lambda(x) = 2^{j/2}\, \psi^{(1)}(2^j x - k),\ x \in [0,1], \qquad \|\psi^{(1)}_\lambda\|_{0,(0,1)} = 1.$$

Then, (1.26) gives an estimate of $\|\psi^{(1)}_\lambda\|_{1,(0,1)}$ by the term 2^j, whereas (2.13) results in view of Theorem 6 in estimate by $K_j := \sqrt{1 + 4^2\, 2^{2j}}$. On the other hand, since $\|\psi^{(0)}_\lambda\|_{0,(0,1)} = 1$, again Theorem 6 implies

$$\|\psi^{(1)}_\lambda\|^2_{1,(0,1)} = \|\psi^{(1)}_\lambda\|^2_{0,(0,1)} + \|4\, 2^j\, \psi^{(0)}_\lambda\|^2_{0,(0,1)} = 1 + 2^{2(j+2)},$$

so that in this case $K_j = \|\psi^{(1)}_\lambda\|_{1,(0,1)}$, i.e., we obtain an *exact* estimate. □

2.2 The Spaces $H(\mathrm{div})$ and $H(\mathrm{curl})$

In this section, we set up our notation and review the basic facts on the spaces of vector fields under consideration that are needed here. We follow the description in [77].

2.2.1 Stream Function Spaces

Let $\Omega \subset \mathbb{R}^n$ be some open bounded domain with Lipschitz-continuous boundary. Then, for $\phi \in L_2(\Omega)$ and a two-dimensional vector field $\boldsymbol{v} = (v_1, v_2)^T$, we define

$$\mathbf{curl}\,\phi := (\partial_2\phi, -\partial_1\phi)^T, \qquad \text{curl}\,\boldsymbol{v} := \partial_1 v_2 - \partial_2 v_1 =: \text{rot}\,\boldsymbol{v}. \tag{2.16}$$

Again we use the short hand notation $\partial_i := \frac{\partial}{\partial x_i}$ for $1 \leq i \leq n$ and we assume the partial derivatives in (2.16) to exist in the weak sense in $L_2(\Omega)$. For three-dimensional vector fields $\boldsymbol{\zeta} = (\zeta_1, \zeta_2, \zeta_3)^T$, we set

$$\mathbf{curl}\,\boldsymbol{\zeta} := (\partial_2\zeta_3 - \partial_3\zeta_2, \partial_3\zeta_1 - \partial_1\zeta_3, \partial_1\zeta_2 - \partial_2\zeta_1)^T, \tag{2.17}$$

where again the partial derivatives should exist in the weak sense in $L_2(\Omega)$.

With these definitions at hand, we define the following stream function spaces (for $n = 3$):

$$\boldsymbol{H}(\mathbf{curl};\Omega) := \{\boldsymbol{\zeta} \in \boldsymbol{L}_2(\Omega) : \mathbf{curl}\,\boldsymbol{\zeta} \in \boldsymbol{L}_2(\Omega)\}, \tag{2.18}$$
$$\boldsymbol{V}(\mathbf{curl};\Omega) := \{\boldsymbol{\zeta} \in \boldsymbol{H}(\mathbf{curl};\Omega) : \mathbf{curl}\,\boldsymbol{\zeta} = \mathbf{0}\}. \tag{2.19}$$

We will be mainly concerned with the 3D-case here. When nothing else is said, $n = 3$ is assumed. We will treat the curl-spaces for $n = 2$ separately. All these spaces are Hilbert spaces with the corresponding graph norm

$$\|\boldsymbol{\zeta}\|^2_{\boldsymbol{H}(\mathbf{curl};\Omega)} := \|\boldsymbol{\zeta}\|^2_{0,\Omega} + \|\mathbf{curl}\,\boldsymbol{\zeta}\|^2_{0,\Omega}. \tag{2.20}$$

Finally, we define

$$\boldsymbol{H}_0(\mathbf{curl};\Omega) := \text{clos}_{\boldsymbol{H}(\mathbf{curl};\Omega)} C_0^\infty(\Omega), \tag{2.21}$$

and

$$\boldsymbol{V}_0(\mathbf{curl};\Omega) := \boldsymbol{H}_0(\mathbf{curl};\Omega) \cap \boldsymbol{V}(\mathbf{curl};\Omega). \tag{2.22}$$

Since these spaces model the stream function in the Navier-Stokes equations for non-primitive variables, we also call them *stream function spaces.*

2.2.2 Flux Spaces

It is known that the stream function spaces are closely linked to a second class of spaces induced by the divergence operator. Since these spaces often model the flux of some physical quantity, we follow [86] and call these spaces *flux spaces.* As usual, the divergence operator is defined for any vector field $\boldsymbol{\zeta} = (\zeta_1, \ldots, \zeta_n)^T$ by

$$\text{div}\,\boldsymbol{\zeta} := \sum_{i=1}^{n} \partial_i\,\zeta_i, \tag{2.23}$$

where it is again assumed that the partial derivatives are well defined in the weak sense in $L_2(\Omega)$. Then, we define

$$\begin{aligned} \boldsymbol{H}(\mathrm{div};\Omega) &:= \{\boldsymbol{\zeta} \in \boldsymbol{L}_2(\Omega) : \mathrm{div}\,\boldsymbol{\zeta} \in L_2(\Omega)\}, \\ \boldsymbol{V}(\mathrm{div};\Omega) &:= \{\boldsymbol{\zeta} \in \boldsymbol{H}(\mathrm{div};\Omega) : \mathrm{div}\,\boldsymbol{\zeta} = 0\}. \end{aligned}$$

Also these spaces are Hilbert spaces under the norm

$$\|\boldsymbol{\zeta}\|^2_{\boldsymbol{H}(\mathrm{div};\Omega)} := \|\boldsymbol{\zeta}\|^2_{0,\Omega} + \|\mathrm{div}\,\boldsymbol{\zeta}\|^2_{0,\Omega}, \tag{2.24}$$

and we define

$$\boldsymbol{H}_0(\mathrm{div};\Omega) := \mathrm{clos}_{\boldsymbol{H}(\mathrm{div};\Omega)} \boldsymbol{C}_0^\infty(\Omega), \tag{2.25}$$

as well as

$$\boldsymbol{V}_0(\mathrm{div};\Omega) := \boldsymbol{H}_0(\mathrm{div};\Omega) \cap \boldsymbol{V}(\mathrm{div};\Omega). \tag{2.26}$$

2.2.3 Hodge Decompositions

There are some well-known relationships between stream function and flux spaces, which we will review now. Especially for the analysis and numerical treatment of partial differential equations involving div- and **curl**-operators, one is interested in the following orthogonal decompositions

$$\begin{aligned} \boldsymbol{L}_2(\Omega) &= \boldsymbol{V}_0(\mathrm{div};\Omega) \oplus^\perp \boldsymbol{V}_0(\mathrm{div};\Omega)^\perp && (2.27)\\ &= \boldsymbol{V}_0(\mathbf{curl};\Omega) \oplus^\perp \boldsymbol{V}_0(\mathbf{curl};\Omega)^\perp, && (2.28) \end{aligned}$$

which are often referred to as *Hodge decompositions*. Both decompositions are also of great interest for replacing $\boldsymbol{L}_2(\Omega)$ by $\boldsymbol{H}_0(\mathrm{div};\Omega)$, $\boldsymbol{H}_0(\mathbf{curl};\Omega)$, respectively.

For (2.27), it is known that

$$\boldsymbol{V}_0(\mathrm{div};\Omega)^\perp \;=\; \{\mathbf{grad}\, q : q \in H^1(\Omega)\} \;=\; \boldsymbol{V}(\mathbf{curl};\Omega) \tag{2.29}$$

for simply connected bounded Lipschitz domains Ω (see [77]). The second decomposition (2.28) is somewhat more involved, one has (see for example [77], p. 50) that

$$\boldsymbol{V}_0(\mathbf{curl};\Omega)^\perp = \{\boldsymbol{\zeta} \in \boldsymbol{H}_0(\mathbf{curl};\Omega) : (\boldsymbol{\zeta}, \mathbf{grad}\,\varphi)_{0,\Omega} = 0, \varphi \in H_0^1(\Omega)\}. \tag{2.30}$$

Moreover, also the fact that $\boldsymbol{V}_0(\mathbf{curl};\Omega)^\perp$ is isomorphic to the space $\boldsymbol{H}^1(\Omega) \cap \boldsymbol{V}(\mathrm{div};\Omega) \cap \boldsymbol{H}_0(\mathbf{curl};\Omega)$ can be found in [77].

2.3 Wavelet Systems for $H(\mathrm{curl})$

Before going into the detailed description of our construction let us sketch the general idea which we illustrate in Fig. 2.4 for $n = 3$. Starting with $\Psi^{(1,1,1)} =$

$\Psi^{(1)}$, we can apply the operators **grad**, **curl** and div in the indicated way. As long as $\Psi^{(0,0,0)} = \Psi^{(0)}$ is a Riesz basis for $L_2(\Omega)$, one can expect that $\boldsymbol{\Psi}^{\mathrm{div}}$ is one for $\boldsymbol{H}(\mathrm{div};\Omega)$ and $\boldsymbol{\Psi}^{\mathrm{curl}}$ one for $\boldsymbol{H}(\mathbf{curl};\Omega)$. This is in fact the main idea behind the construction, which is similar e.g. to Nédélec finite element spaces for $\boldsymbol{H}(\mathrm{div};\Omega)$ and $\boldsymbol{H}(\mathbf{curl};\Omega)$, see, e.g. [86].

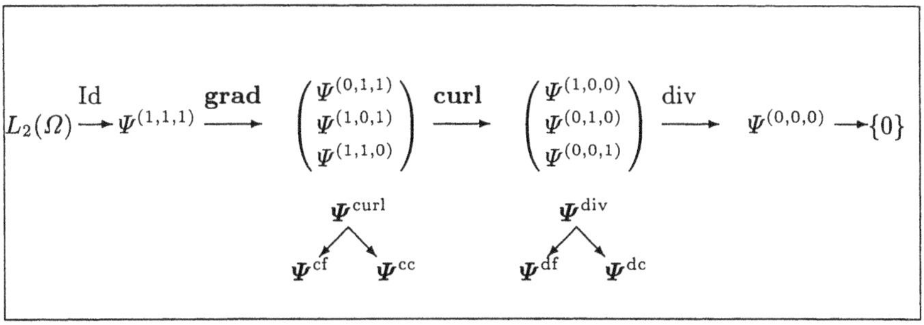

Fig. 2.4. Application of differential operators to special wavelet bases. The arrows indicate, e.g., that $\mathbf{grad}(S(\Psi^{(1,1,1)})) = S(\boldsymbol{\Psi}^{\mathrm{curl}})$, $\mathbf{curl}(S(\boldsymbol{\Psi}^{\mathrm{curl}})) = S(\boldsymbol{\Psi}^{\mathrm{div}})$ and $\mathrm{div}(S(\boldsymbol{\Psi}^{\mathrm{div}})) = S(\Psi^{(0,0,0)})$. The series is as also *exact*, [86].
It has to be understood that $\boldsymbol{\Psi}^{\mathrm{curl}} = (\Psi^{(0,1,1)}\boldsymbol{\delta}_1) \oplus (\Psi^{(1,0,1)}\boldsymbol{\delta}_2) \oplus (\Psi^{(1,1,0)}\boldsymbol{\delta}_3)$ as well as $\boldsymbol{\Psi}^{\mathrm{div}} = (\Psi^{(1,0,0)}\boldsymbol{\delta}_1) \oplus (\Psi^{(0,1,0)}\boldsymbol{\delta}_2) \oplus (\Psi^{(0,0,1)}\boldsymbol{\delta}_3)$.

2.3.1 Wavelets in $\boldsymbol{H}_0(\mathbf{curl};\Omega)$

We start defining a wavelet basis for the space $\boldsymbol{H}_0(\mathbf{curl};\Omega)$ in 3D. The definition is also valid for the 2D-case, where one has to consider the space $\boldsymbol{H}_0(\mathrm{curl};\Omega)^2$ defined in an obvious fashion.

Definition 3. *Let* $\mathbf{1} := (1,\dots,1)^T \in E^n$ *and define vector wavelet functions for* $\boldsymbol{\lambda} = (i,\lambda) \in \boldsymbol{\mathcal{J}} := \{(i,\lambda) : i = 1,\dots,n; \lambda \in \mathcal{J}\}$ *by*

$$\psi_{\boldsymbol{\lambda}}^{\mathrm{curl}} := \psi_\lambda^{(\mathbf{1}-\boldsymbol{\delta}_i)}\,\boldsymbol{\delta}_i, \qquad \tilde{\psi}_{\boldsymbol{\lambda}}^{\mathrm{curl}} := \tilde{\psi}_\lambda^{(\mathbf{1}-\boldsymbol{\delta}_i)}\,\boldsymbol{\delta}_i. \tag{2.31}$$

Accordingly, we define $\boldsymbol{\Psi}^{\mathrm{curl}} := \{\psi_{\boldsymbol{\lambda}}^{\mathrm{curl}} : \boldsymbol{\lambda} \in \boldsymbol{\mathcal{J}}\}$ *and* $\tilde{\boldsymbol{\Psi}}^{\mathrm{curl}} := \{\tilde{\psi}_{\boldsymbol{\lambda}}^{\mathrm{curl}} : \boldsymbol{\lambda} \in \boldsymbol{\mathcal{J}}\}$. □

For the above defined basis, we want to derive a norm equivalence that allows to estimate the $\|\cdot\|_{\boldsymbol{H}(\mathbf{curl};\Omega)}$-norm of each function $\boldsymbol{\zeta} \in \boldsymbol{H}(\mathbf{curl};\Omega)$ by a (discrete) norm of the coefficients in the expansion of $\boldsymbol{\zeta}$ in terms of $\boldsymbol{\Psi}^{\mathrm{curl}}$. Let us now define this norm, where we restrict ourselves to the 3D-case, while a similar definition applies to $n = 2$. For a given sequence $\boldsymbol{c} \in \ell(\boldsymbol{\mathcal{J}})$, we define

$$\mathsf{curl}\,\boldsymbol{c} := (\mathsf{d}^{(2)}\,c^{(3)} - \mathsf{d}^{(3)}\,c^{(2)}, \mathsf{d}^{(3)}\,c^{(1)} - \mathsf{d}^{(1)}\,c^{(3)}, \mathsf{d}^{(1)}\,c^{(2)} - \mathsf{d}^{(2)}\,c^{(1)})^T, \tag{2.32}$$

where we have used the short hand notation $c^{(i)} := \{c_{(i,\lambda)}\}_{\lambda\in\mathcal{J}}$. Then, we set

$$\|\mathbf{c}\|_{h(\mathsf{curl};\mathcal{J})} := (\|\mathbf{c}\|^2_{\ell_2(\mathcal{J})} + \|\mathsf{curl}\,\mathbf{c}\|^2_{\ell_2(\mathcal{J})})^{1/2}. \tag{2.33}$$

We obtain the following result.

Theorem 8. *Under Ass. 3, we have for $n = 2,3$ that the systems $\boldsymbol{\Psi}^{\mathrm{curl}}$, $\tilde{\boldsymbol{\Psi}}^{\mathrm{curl}}$ form a biorthogonal generalized wavelet basis in the sense of Definition 1 on p. 2 for $\boldsymbol{H}_0(\mathbf{curl};\Omega)$. Moreover, for $\boldsymbol{\zeta} \in \boldsymbol{H}_0(\mathbf{curl};\Omega) = \sum_{\boldsymbol{\lambda}\in\mathcal{J}} c_{\boldsymbol{\lambda}}\,\boldsymbol{\psi}^{\mathrm{curl}}_{\boldsymbol{\lambda}}$ we have*

$$\|\boldsymbol{\zeta}\|_{\boldsymbol{H}(\mathbf{curl};\Omega)} \sim \|\mathbf{c}\|_{h(\mathsf{curl};\mathcal{J})}. \tag{2.34}$$

Proof. We will only give the proof for $n = 3$ and remark that the case $n = 2$ can be treated completely analogously. Let us first check that the functions $\boldsymbol{\psi}^{\mathrm{curl}}_{\boldsymbol{\lambda}}$, $\boldsymbol{\lambda} \in \mathcal{J}$, are indeed in $\boldsymbol{H}(\mathbf{curl};\Omega)$. To this end, set again $\boldsymbol{\lambda} = (i,\lambda) \in \mathcal{J}$. Then, we have $[\mathbf{curl}\,\boldsymbol{\psi}^{\mathrm{curl}}_{\boldsymbol{\lambda}}]_i = 0$ which is trivially in $L_2(\Omega)$. For $i \neq i'$ we obtain $\partial_{i'}[\boldsymbol{\psi}^{\mathrm{curl}}_{\boldsymbol{\lambda}}]_i = [D_{|\lambda|}]^{\lambda}\,\Psi_j^{(\mathbf{1}-\boldsymbol{\delta}_i-\boldsymbol{\delta}_{i'})}$ which, by assumption, is a function in $L_2(\Omega)$ and hence $\mathbf{curl}\,\boldsymbol{\psi}^{\mathrm{curl}}_{\boldsymbol{\lambda}} \in \boldsymbol{L}_2(\Omega)$.

The biorthogonality is readily seen:

$$(\boldsymbol{\psi}^{\mathrm{curl}}_{\boldsymbol{\lambda}}, \tilde{\boldsymbol{\psi}}^{\mathrm{curl}}_{\boldsymbol{\lambda}'})_{\boldsymbol{L}_2(\Omega)} = \delta_{i,i'}\,(\psi_\lambda^{(\mathbf{1}-\boldsymbol{\delta}_i)}, \tilde{\psi}_{\lambda'}^{(\mathbf{1}-\boldsymbol{\delta}_i)})_{L_2(\Omega)} = \delta_{i,i'}\,\delta_{\lambda,\lambda'} = \delta_{\boldsymbol{\lambda},\boldsymbol{\lambda}'},$$

for all $\boldsymbol{\lambda}, \boldsymbol{\lambda}' \in \mathcal{J}$.

Next, we have to show that each function $\boldsymbol{\zeta} \in \boldsymbol{H}_0(\mathbf{curl};\Omega)$ has a unique expansion in terms of $\boldsymbol{\Psi}^{\mathrm{curl}}$. By assumption, each $\Psi^{(\mathbf{1}-\boldsymbol{\delta}_i)}$, $1 \le i \le n$, forms a Riesz basis for $L_2(\Omega)$. Since $\boldsymbol{\zeta} \in \boldsymbol{H}_0(\mathbf{curl};\Omega) \subset \boldsymbol{L}_2(\Omega)$, this function can in fact uniquely be expanded in terms of $\boldsymbol{\Psi}^{\mathrm{curl}}$.

Finally, we have to prove (2.34). The norm equivalences for $\Psi^{(\mathbf{1}-\boldsymbol{\delta}_i)}$ imply that $\|\boldsymbol{\zeta}\|_{\boldsymbol{L}_2(\Omega)} \sim \|\mathbf{c}\|_{\ell_2(\mathcal{J})}$. By (2.15), we obtain

$$(\mathbf{curl}\,\boldsymbol{\zeta})_i = \sum_{\lambda\in\mathcal{J}} (\mathsf{curl}\,\mathbf{c})_{(i,\lambda)}\,\psi_\lambda^{(\mathbf{0})}, \quad 1 \le i \le n.$$

Using the norm equivalences for $\Psi^{(\mathbf{0})}$ leads to

$$\|\mathbf{curl}\,\boldsymbol{\zeta}\|^2_{\boldsymbol{L}_2(\Omega)} \sim \sum_{i=1}^{3}\sum_{\lambda\in\mathcal{J}} |(\mathsf{curl}\,\mathbf{c})_{(i,\lambda)}|^2 = \|\mathsf{curl}\,\mathbf{c}\|^2_{\ell_2(\mathcal{J})}, \tag{2.35}$$

so that we obtain (2.34). □

For later purposes, it is important to consider also weightened norms.

Corollary 2. *Under the assumptions of Theorem 8, we have for $\boldsymbol{\zeta} = \mathbf{c}^T\,\boldsymbol{\Psi}^{\mathrm{curl}}$ and ν, $\mu \in \mathbb{R}$*

$$\nu\|\boldsymbol{\zeta}\|^2_{0,\Omega} + \mu\|\mathbf{curl}\,\boldsymbol{\zeta}\|^2_{0,\Omega} \sim \nu\|\mathbf{c}\|^2_{\ell_2(\mathcal{J})} + \mu\|\mathsf{curl}\,\mathbf{c}\|^2_{\ell_2(\mathcal{J})}.$$

Proof. In the proof of Theorem 8 it was shown that $\|\boldsymbol{\zeta}\|_{0,\Omega} \sim \|\mathbf{c}\|_{\ell_2(\mathcal{J})}$ and $\|\mathbf{curl}\,\boldsymbol{\zeta}\|_{0,\Omega} \sim \|\mathsf{curl}\,\mathbf{c}\|_{\ell_2(\mathcal{J})}$ which already proves the assertion. □

Remark 5. The estimate (2.34) shows that $\mathbf{curl}\,\boldsymbol{\zeta} = 0$ in $\boldsymbol{L}_2(\Omega)$ if and only if $\mathsf{curl}\,\boldsymbol{c} \equiv 0$. Hence, given a wavelet expansion of $\boldsymbol{\zeta} \in \boldsymbol{H}(\mathbf{curl};\Omega)$ in terms of $\boldsymbol{\Psi}^{\mathrm{curl}}$, we can check whether $\boldsymbol{\zeta}$ is **curl**-free or not by only considering the wavelet coefficients. This will turn out to be very useful in the sequel. Moreover, (2.34) implies for $\boldsymbol{\zeta} \in \boldsymbol{L}_2(\Omega)$ that $\boldsymbol{\zeta} \in \boldsymbol{H}(\mathbf{curl};\Omega)$ if and only if $\mathsf{curl}\,\boldsymbol{c} \in \ell_2(\mathcal{J})$. □

Remark 6. When skipping hypothesis (b) of Ass. 3, it is also possible to study wavelet bases in $\boldsymbol{H}(\mathbf{curl};\Omega)$ instead of $\boldsymbol{H}_0(\mathbf{curl};\Omega)$. The above theorem remains valid for this case, provided that the 1D systems are equipped with the appropriate boundary conditions for $\boldsymbol{H}_0(\mathbf{curl};\Omega)$, [17]. □

Remark 7. For later purposes, we note the following estimate

$$\Big\| \sum_{\boldsymbol{\lambda}\in\mathcal{J}} c_{\boldsymbol{\lambda}}\, \boldsymbol{\psi}_{\boldsymbol{\lambda}}^{\mathrm{curl}} \Big\|_{s,\Omega}^2 \sim \sum_{\boldsymbol{\lambda}\in\mathcal{J}} 2^{2|\boldsymbol{\lambda}|s} |c_{\boldsymbol{\lambda}}|^2, \tag{2.36}$$

provided that all systems $\Psi^{(\mathbf{1}-\boldsymbol{\delta}_i)}$, $1 \le i \le n$, give rise to the characterization (1.26) on p. 9 for $m = s \in \mathbb{R}$. In fact, (2.36) is an easy consequence of Definition 3 and (1.26). □

Let us finally mention that also the functions

$$\check{\boldsymbol{\psi}}_{\boldsymbol{\lambda}}^{\mathrm{curl}} := \psi_{\lambda}^{[i]}\, \boldsymbol{\delta}_i, \qquad \check{\tilde{\boldsymbol{\psi}}}_{\boldsymbol{\lambda}}^{\mathrm{curl}} := \tilde{\psi}_{\lambda}^{[i]}\, \boldsymbol{\delta}_i$$

for *any* wavelet systems $\Psi^{[i]}$ in $L_2(\Omega)$, $1 \le i \le n$, give rise to a wavelet basis in $\boldsymbol{H}_0(\mathbf{curl};\Omega)$ provided that the functions $\psi_{\lambda}^{[i]}$ are sufficiently smooth (in order to ensure $\check{\boldsymbol{\psi}}_{\boldsymbol{\lambda}}^{\mathrm{curl}} \in \boldsymbol{H}_0(\mathbf{curl};\Omega)$). However, the corresponding norm equivalence will then in general not take such a nice structure as in (2.34). By 'nice' we mean that $\mathsf{curl}\,\mathbf{c}$ vanishes for $\boldsymbol{\zeta} \in \boldsymbol{V}_0(\mathbf{curl};\Omega)$ (see Remark 5) which is due to the special choice of the basis functions. Such a property cannot be expected by using arbitrary systems. Moreover, it will become clear in the next subsection that this particular choice enables us to construct **curl**-free vector wavelets.

2.3.2 Curl-Free Wavelet Bases

In this subsection we construct **curl**-free wavelet bases, which will be defined as follows. Let us again focus on the 3D-case.

Definition 4. *For $\lambda \in \mathcal{J}$ and $1 \le i \le n$, we define*

$$(\psi_{\lambda}^{\mathrm{cf}})_{|i} := [D_{|\lambda|}^{(i)}]^{\lambda}\, \Psi_{|\lambda|}^{(\mathbf{1}-\boldsymbol{\delta}_i)}, \qquad \tilde{\psi}_{\lambda}^{\mathrm{cf}} := [(D_{i_\lambda}^{(i_\lambda)})^{-1}]_{\lambda}^{T}\, \tilde{\Psi}_{|\lambda|}^{(\mathbf{1}-\boldsymbol{\delta}_{i_\lambda})}\, \boldsymbol{\delta}_{i_\lambda} \tag{2.37}$$

as well as $\boldsymbol{\Psi}^{\mathrm{cf}} := \{\psi_{\lambda}^{\mathrm{cf}} : \lambda \in \mathcal{J}\}$, $\tilde{\boldsymbol{\Psi}}^{\mathrm{cf}} := \{\tilde{\psi}_{\lambda}^{\mathrm{cf}} : \lambda \in \mathcal{J}\}$. □

Note that we have

$$\psi_\lambda^{\mathrm{cf}} = \mathbf{grad}\,\psi_\lambda^{(1)}, \tag{2.38}$$

which is in spirit of (2.29).

Theorem 9. *Under the hypotheses in Ass. 3, the systems of functions $\boldsymbol{\Psi}^{\mathrm{cf}}$, $\tilde{\boldsymbol{\Psi}}^{\mathrm{cf}}$ form a biorthogonal generalized wavelet basis in the sense of Definition 1 on p. 2 for $\boldsymbol{V}_0(\mathbf{curl};\Omega)$ for $n = 2, 3$.*

Proof. It can easily be seen that $\psi_\lambda^{\mathrm{cf}}$ are in fact **curl**-free. Moreover, they are obviously linear combinations of $\boldsymbol{\Psi}^{\mathrm{curl}}$ and $\tilde{\boldsymbol{\Psi}}^{\mathrm{curl}}$, respectively and hence in $\boldsymbol{H}_0(\mathbf{curl};\Omega)$. Using (2.7), integration by parts and the biorthogonality of the scalar systems, we obtain the biorthogonality of the vector fields:

$$\begin{aligned}(\psi_\lambda^{\mathrm{cf}}, \tilde{\psi}_{\lambda'}^{\mathrm{cf}})_{0,\Omega} &= \left(\partial_{i_{\lambda'}}\psi_\lambda^{(1)}, [(D_{i_{\lambda'}}^{(i_{\lambda'})})^{-1}]_{\lambda'}^T \tilde{\Psi}_{|\lambda'|}^{(1-\delta_{i_{\lambda'}})}\right)_{0,\Omega} \\ &= (-1)\left(\psi_\lambda^{(1)}, [(D_{i_{\lambda'}}^{(i_{\lambda'})})^{-1}]_{\lambda'}^T \partial_{i_{\lambda'}}\tilde{\Psi}_{|\lambda'|}^{(1-\delta_{i_{\lambda'}})}\right)_{0,\Omega} \\ &= (\psi_\lambda^{(1)}, \tilde{\psi}_{\lambda'}^{(1)})_{0,\Omega} = \delta_{\lambda,\lambda'}, \quad \lambda, \lambda' \in \mathcal{J}.\end{aligned}$$

Now it remains to prove that the system $\boldsymbol{\Psi}^{\mathrm{cf}}$ is in fact a basis for $\boldsymbol{V}_0(\mathbf{curl};\Omega)$, i.e., that each vector field in $\boldsymbol{V}_0(\mathbf{curl};\Omega)$ has a unique expansion in terms of $\boldsymbol{\Psi}^{\mathrm{cf}}$. To this end, let $\boldsymbol{\zeta} \in \boldsymbol{V}_0(\mathbf{curl};\Omega)$. Since $\boldsymbol{V}_0(\mathbf{curl};\Omega) \subset \boldsymbol{L}_2(\Omega)$, Ass. 3 yields

$$\begin{aligned}\zeta_i &= \sum_{\lambda\in\mathcal{J}} (\zeta_i, \tilde{\psi}_\lambda^{(1-\delta_i)})_{0,\Omega}\, \psi_\lambda^{(1-\delta_i)}, \\ &= \Bigg(\sum_{\substack{\lambda\in\mathcal{J}\\ i_\lambda = i}} + \sum_{\substack{\lambda\in\mathcal{J}\\ i_\lambda \neq i}}\Bigg)(\zeta_i, \tilde{\psi}_\lambda^{(1-\delta_i)})_{0,\Omega}\, \psi_\lambda^{(1-\delta_i)}\end{aligned} \tag{2.39}$$

for $1 \le i \le n$. Let us first consider the case $i = i_\lambda$. By our assumption, we have $\partial_{i_\lambda}\Psi_{|\lambda|}^{(1)} = D_{i_\lambda}^{(i_\lambda)}\Psi_{|\lambda|}^{(1-\delta_{i_\lambda})}$, which implies

$$\psi_\lambda^{(1-\delta_{i_\lambda})} = [(D_{i_\lambda}^{(i_\lambda)})^{-1}]^\lambda\, \partial_{i_\lambda}\Psi_{|\lambda|}^{(1)}.$$

Hence we obtain for the first sum in (2.39)

$$\begin{aligned}&\sum_{\substack{\lambda\in\mathcal{J}\\ i_\lambda = i}} (\zeta_i, \tilde{\psi}_\lambda^{(1-\delta_i)})_{0,\Omega}\, \psi_\lambda^{(1-\delta_i)} \\ &\quad= \sum_{\substack{\lambda\in\mathcal{J}\\ i_\lambda = i}} (\zeta_{i_\lambda}, \tilde{\psi}_\lambda^{(1-\delta_{i_\lambda})})_{0,\Omega}\, [(D_{i_\lambda}^{(i_\lambda)})^{-1}]^\lambda\, \partial_{i_\lambda}\Psi_{|\lambda|}^{(1)} \\ &\quad= \sum_{\substack{\lambda\in\mathcal{J}\\ i_\lambda = i}} \left(\zeta_{i_\lambda}, [(D_{i_\lambda}^{(i_\lambda)})^{-1}]_\lambda^T \tilde{\Psi}_{|\lambda|}^{(1-\delta_{i_\lambda})}\right)_{0,\Omega} \partial_{i_\lambda}\Psi_{|\lambda|}^{(1)},\end{aligned} \tag{2.40}$$

where in the last equation we have used summation by parts in the sense

$$\sum_{\lambda\in\Lambda}(f,g_\lambda)_{0,\Omega}\,[A]^\lambda v = \sum_{\lambda,\mu\in\Lambda}(f,g_\lambda)_{0,\Omega}\,a_{\lambda,\mu}v_\mu$$
$$= \sum_{\mu\in\Lambda}\Big(f,\sum_{\lambda\in\Lambda}a_{\lambda,\mu}\,g_\lambda\Big)_{0,\Omega}v_\mu \;=\; \sum_{\lambda\in\Lambda}(f,[A]^T_\lambda g)_{0,\Omega}\,v_\lambda$$

for each countable set Λ, $v=(v_\lambda)_{\lambda\in\Lambda}$, $g=(g_\lambda)_{\lambda\in\Lambda}$ and $A=(a_{\lambda,\mu})_{\lambda,\mu\in\Lambda}$.

For the second term in (2.39), we use the fact that $\partial_i\zeta_{i_\lambda}=\partial_{i_\lambda}\zeta_i$ for $i\neq i_\lambda$, which is a consequence of $\mathbf{curl}\,\boldsymbol{\zeta}=0$. Moreover, by our assumption, we have

$$\tilde{\psi}_\lambda^{(1-\delta_i)} \;=\; -\,[(D_{i_\lambda}^{(i_\lambda)})^{-1}]^T_\lambda\;\partial\tilde{\Psi}_{|\lambda|}^{(1\delta_i-\delta_{i_\lambda})} \;=\; (-1)\partial_{i_\lambda}[(D_{i_\lambda}^{(i_\lambda)})^{-1}]^T_\lambda\;\tilde{\Psi}_{|\lambda|}^{(1\delta_i-\delta_{i_\lambda})},$$

so that by integration by parts, we obtain for $i\neq i_\lambda$

$$(\zeta_i,\tilde{\psi}_\lambda^{(1-\delta_i)})_{0,\Omega} = (-1)[(D_{i_\lambda}^{(i_\lambda)})^{-1}]^T_\lambda\;(\zeta_i,\partial_{i_\lambda}\tilde{\Psi}_{|\lambda|}^{(1-\delta_i-\delta_{i_\lambda})})_{0,\Omega}$$
$$= (-1)[(D_{i_\lambda}^{(i_\lambda)})^{-1}]^T_\lambda\;(\zeta_{i_\lambda},\partial_i\tilde{\Psi}_{|\lambda|}^{(1-\delta_i-\delta_{i_\lambda})})_{0,\Omega}$$
$$= [(D_{i_\lambda}^{(i_\lambda)})^{-1}]^T_\lambda\;(\zeta_{i_\lambda},\tilde{\Psi}_{|\lambda|}^{(1-\delta_{i_\lambda})})_{0,\Omega}\;(D_i^{(i)})^T,$$

which, in turn, implies again using summation by parts

$$\sum_{\substack{\lambda\in\mathcal{J}\\ i_\lambda\neq i}}(\zeta_i,\tilde{\psi}_\lambda^{(1-\delta_i)})_{0,\Omega}\,\psi_\lambda^{(1-\delta_i)}$$
$$= \sum_{\substack{\lambda\in\mathcal{J}\\ i_\lambda\neq i}}(\zeta_{i_\lambda},[(D_{i_\lambda}^{(i_\lambda)})^{-1}]^T_\lambda\,\tilde{\Psi}_{|\lambda|}^{(1-\delta_{i_\lambda})})_{0,\Omega}\,[D_i^{(i)}]^\lambda\,\Psi_{|\lambda|}^{(1-\delta_i)}$$
$$= \sum_{\substack{\lambda\in\mathcal{J}\\ i_\lambda\neq i}}(\zeta_{i_\lambda},[(D_{i_\lambda}^{(i_\lambda)})^{-1}]^T_\lambda\,\tilde{\Psi}_{|\lambda|}^{(1-\delta_{i_\lambda})})_{0,\Omega}\,\partial_i\psi_\lambda^{(1)}.$$

Finally, using the latter equation, (2.39) and (2.40) yields

$$\zeta_i=\sum_{\lambda\in\mathcal{J}}(\zeta_{i_\lambda},[(D_{i_\lambda}^{(i_\lambda)})^{-1}]^T_\lambda\,\tilde{\Psi}_{|\lambda|}^{(1-\delta_{i_\lambda})})_{0,\Omega}\,\partial_i\psi_\lambda^{(1)} = \sum_{\lambda\in\mathcal{J}}(\boldsymbol{\zeta},\tilde{\boldsymbol{\psi}}_\lambda^{\mathrm{cf}})_{0,\Omega}\,(\boldsymbol{\psi}_\lambda^{\mathrm{cf}})_{|i}$$

for $1\le i\le n$, which completes the proof. □

The above result also gives rise to a characterization of $\boldsymbol{V}_0(\mathbf{curl};\Omega)'$. Of course, since the space $\boldsymbol{V}_0(\mathbf{curl};\Omega)$ is a closed subspace of $\boldsymbol{L}_2(\Omega)$ (see e.g. [77]), the dual space of $\boldsymbol{V}_0(\mathbf{curl};\Omega)$ (as all Hilbert spaces) is isomorphic to the space itself. However, it may be helpful to have a concrete representation at hand. We would like to point out first that the dual system $\tilde{\boldsymbol{\Psi}}^{\mathrm{cf}}$ is in general *not* in $\boldsymbol{V}_0(\mathbf{curl};\Omega)$. The biorthogonality between $\boldsymbol{\Psi}^{\mathrm{cf}}$ and $\tilde{\boldsymbol{\Psi}}^{\mathrm{cf}}$ however ensures that $\tilde{\boldsymbol{\Psi}}^{\mathrm{cf}}\subset\boldsymbol{V}_0(\mathbf{curl};\Omega)'$. In view of Theorem 9, this already proves the following result.

Corollary 3. *A vector field* $\boldsymbol{\zeta} \in \boldsymbol{L}_2(\Omega)$ *is in* $\mathbf{V}_0(\mathbf{curl};\Omega)'$ *if and only if*

$$\sum_{\lambda \in \mathcal{J}} |(\boldsymbol{\zeta}, \boldsymbol{\psi}_\lambda^{\mathrm{cf}})_{0,\Omega}|^2 < \infty. \qquad \square \tag{2.41}$$

Finally, we derive some norm equivalences. If Ω is a bounded domain, (2.38) and the Poincaré-Friedrichs inequality imply the estimate for $\Psi^{(1)} \subset H_0^1(\Omega)$

$$\left\| \sum_{\lambda \in \mathcal{J}} c_\lambda \boldsymbol{\psi}_\lambda^{\mathrm{cf}} \right\|_{s,\Omega}^2 \sim \sum_{\lambda \in \mathcal{J}} 2^{2(s+1)|\lambda|} |c_\lambda|^2, \tag{2.42}$$

provided that the norm equivalence (1.26) holds for $\Psi^{(1)}$ and $m = s+1$. Again, we obtain another estimate using the operator grad defined in (2.14) on p. 89.

Proposition 4. *Under Ass. 3, we have*

$$\left\| \sum_{\lambda \in \mathcal{J}} c_\lambda \boldsymbol{\psi}_\lambda^{\mathrm{cf}} \right\|_{0,\Omega} \sim \|\mathsf{grad}\, \mathbf{c}\|_{\ell_2(\mathcal{J})}. \qquad \square$$

2.4 Wavelet Bases for $\boldsymbol{H}(\mathrm{div})$

Now, we consider the flux spaces and construct appropriate wavelet bases. We will show the interrelations to the wavelet bases in $\boldsymbol{H}(\mathbf{curl};\Omega)$. These or similar constructions can also be found in [52, 98, 116, 117, 118].

2.4.1 Wavelet Bases in $\boldsymbol{H}(\mathrm{div};\Omega)$

With the results of Sect. 2.3 at hand, it is not hard to see that the wavelet systems $\boldsymbol{\Psi}^{\mathrm{div}}$, $\tilde{\boldsymbol{\Psi}}^{\mathrm{div}}$ defined by the functions

$$\boldsymbol{\psi}_{\boldsymbol{\lambda}}^{\mathrm{div}} := \psi_\lambda^{(\boldsymbol{\delta}_i)}\, \boldsymbol{\delta}_i, \quad \tilde{\boldsymbol{\psi}}_{\boldsymbol{\lambda}}^{\mathrm{div}} := \tilde{\psi}_\lambda^{(\boldsymbol{\delta}_i)}\, \boldsymbol{\delta}_i, \qquad \boldsymbol{\lambda} = (i, \lambda) \in \boldsymbol{\mathcal{J}}, \tag{2.43}$$

form a biorthogonal system in $\boldsymbol{H}(\mathrm{div};\Omega)$. Moreover, the norm equivalence

$$\|\boldsymbol{\zeta}\|_{\boldsymbol{H}(\mathrm{div};\Omega)} \sim \|\mathbf{c}\|_{h(\mathrm{div};\boldsymbol{\mathcal{J}})} \tag{2.44}$$

holds for any $\boldsymbol{\zeta} = \sum_{\boldsymbol{\lambda} \in \boldsymbol{\mathcal{J}}} c_{\boldsymbol{\lambda}} \boldsymbol{\psi}_{\boldsymbol{\lambda}}^{\mathrm{div}} \in \boldsymbol{H}(\mathrm{div};\Omega)$, where we have defined

$$\mathsf{div} : \ell(\boldsymbol{\mathcal{J}}) \to \ell(\mathcal{J}), \qquad \mathsf{div}\, \mathbf{c} := \sum_{i=1}^n \mathsf{d}^{(i)}\, c^{(i)},$$

and

$$\|\boldsymbol{c}\|_{h(\mathrm{div};\boldsymbol{\mathcal{J}})} := (\|\boldsymbol{c}\|^2_{\ell_2(\boldsymbol{\mathcal{J}})} + \|\mathrm{div}\,\boldsymbol{c}\|^2_{\ell_2(\mathcal{J})})^{1/2}. \tag{2.45}$$

In fact, by using a similar reasoning as in (2.15), we obtain

$$\begin{aligned}
\|\zeta\|^2_{\boldsymbol{H}(\mathrm{div};\Omega)} &= \|\zeta\|^2_{0,\Omega} + \left\| \sum_{i=1}^{n} \sum_{\lambda\in\mathcal{J}} c_{\boldsymbol{\lambda}}\, \partial_i \psi_\lambda^{(\delta_i)} \right\|^2_{0,\Omega} \\
&= \|\zeta\|^2_{0,\Omega} + \left\| \sum_{\lambda\in\mathcal{J}} \sum_{i=1}^{n} (\mathrm{d}^{(i)} c^{(i)})_\lambda\, \psi_\lambda^{(0)} \right\|^2_{0,\Omega} \\
&\sim \|\boldsymbol{c}\|^2_{\ell_2(\boldsymbol{\mathcal{J}})} + \sum_{\lambda\in\mathcal{J}} |(\mathrm{div}\,\boldsymbol{c})_\lambda|^2 \;=\; \|\boldsymbol{c}\|^2_{h(\mathrm{div};\boldsymbol{\mathcal{J}})},
\end{aligned} \tag{2.46}$$

which proves (2.44). In a similar way we obtain

$$\nu\|\zeta\|^2_{0,\Omega} + \mu\|\mathrm{div}\,\zeta\|^2_{0,\Omega} \sim \nu\|\boldsymbol{c}\|^2_{\ell_2(\boldsymbol{\mathcal{J}})} + \mu\|\mathrm{div}\,\boldsymbol{c}\|^2_{\ell_2(\mathcal{J})}.$$

This shows that Ψ^{div}, $\tilde{\Psi}^{\mathrm{div}}$ form a biorthogonal generalized wavelet basis of $\boldsymbol{H}(\mathrm{div};\Omega)$ and $\boldsymbol{H}(\mathrm{div};\Omega)'$, respectively, in the sense of Definition 1 on p. 2. Finally, by using the standard estimate (1.26) on p. 9, we obtain

$$\|\zeta\|^2_{s,\Omega} \sim \sum_{\boldsymbol{\lambda}\in\boldsymbol{\mathcal{J}}} 2^{2|\boldsymbol{\lambda}|s}\, |c_{\boldsymbol{\lambda}}|^2 \tag{2.47}$$

provided that all $\Psi^{(\delta_i)}$, $1 \le i \le n$, fulfill (1.26) for $m = s$.

2.4.2 Divergence-Free Wavelet Bases

We recall the construction of divergence-free wavelets. The construction of compactly supported divergence-free wavelets was initiated by P.G. Lemarié-Rieusset [98] (see also [116] for the generalization to non-tensor product functions).

Definition 5. *For*

$$\boldsymbol{\lambda} \in \boldsymbol{\mathcal{J}}^{\mathrm{df}} := \{\boldsymbol{\lambda} = (i,\lambda) : \lambda \in \mathcal{J},\, 1 \le i \le n,\, i \ne i_\lambda\},$$

we define

$$(\psi^{\mathrm{df}}_{\boldsymbol{\lambda}})_{|\nu} := \begin{cases} 0, & \nu \notin \{i, i_\lambda\}, \\ \psi_\lambda^{(\delta_i)}, & \nu = i, \\ -\partial_i (D_{i_\lambda}^{(i_\lambda)})^{-1} \Psi_{|\lambda|}^{(\delta_i+\delta_{i_\lambda})}, & \nu = i_\lambda. \end{cases} \tag{2.48}$$

as well as

$$\tilde{\psi}^{\mathrm{df}}_{\boldsymbol{\lambda}} := \tilde{\psi}_\lambda^{(\delta_i)}\, \delta_i, \quad \boldsymbol{\lambda} \in \boldsymbol{\mathcal{J}}^{\mathrm{df}}. \tag{2.49}$$

The systems Ψ^{df} and $\tilde{\Psi}^{\mathrm{df}}$ are defined in a straightforward way. □

For convenience, let us briefly check that these functions are divergence-free. In fact, using (2.7) on p. 87 again, we obtain:

$$\begin{aligned}\operatorname{div} \boldsymbol{\psi}_{\boldsymbol{\lambda}}^{\mathrm{df}} &= \partial_i \psi_{\lambda}^{(\delta_i)} - \partial_{i_\lambda} \partial_i (D_{i_\lambda}^{(i_\lambda)})^{-1} \Psi_{|\lambda|}^{(\delta_i+\delta_{i_\lambda})} \\ &= \partial_{i_\lambda} \partial_i (D_{i_\lambda}^{(i_\lambda)})^{-1} \Psi_{|\lambda|}^{(\delta_i+\delta_{i_\lambda})} - \partial_{i_\lambda} \partial_i (D_{i_\lambda}^{(i_\lambda)})^{-1} \Psi_{|\lambda|}^{(\delta_i+\delta_{i_\lambda})} = 0.\end{aligned}$$

In summary, we have

Theorem 10. *[98, 116, 117] Under Ass. 3 the following statements hold:*

(a) $\boldsymbol{\Psi}^{\mathrm{div}}$, $\tilde{\boldsymbol{\Psi}}^{\mathrm{div}}$ *form a biorthogonal system for* $\boldsymbol{H}_0(\mathrm{div};\Omega)$.
(b) $\boldsymbol{\Psi}^{\mathrm{df}}$ *forms a Riesz basis for* $\boldsymbol{V}_0(\mathrm{div};\Omega)$.

This means in particular, that the above systems form generalized wavelet bases in the sense of Definition 1 on p. 2 for the according spaces. □

Finally, we aim at deriving a norm equivalence in $\boldsymbol{H}^s(\Omega)$. Since this result is not included in the above mentioned references, we will give it in detail. Let us consider a divergence-free vector field

$$\zeta = \sum_{\boldsymbol{\lambda}\in\mathcal{J}^{\mathrm{df}}} c_{\boldsymbol{\lambda}} \boldsymbol{\psi}_{\boldsymbol{\lambda}}^{\mathrm{df}}, \qquad \boldsymbol{c} \in \ell_2(\mathcal{J}^{\mathrm{df}}). \tag{2.50}$$

Proposition 5. *Under Ass. 3, the following equivalence holds for the vector field* ζ *in* (2.50)

$$\|\zeta\|_{s,\Omega}^2 \sim \sum_{\boldsymbol{\lambda}\in\mathcal{J}^{\mathrm{df}}} 2^{2s|\boldsymbol{\lambda}|} |c_{\boldsymbol{\lambda}}|^2. \qquad \square$$

2.5 Helmholtz and Hodge Decompositions

The wavelet bases above described give rise to Helmholtz and Hodge decompositions as introduced in Sect. 2.2.3. We will investigate now these decompositions in terms of the wavelet bases.

2.5.1 A Biorthogonal Helmholtz Decomposition

The wavelets defined in (2.48) give rise to a Helmholtz decomposition of the space $\boldsymbol{H}(\mathrm{div};\Omega)$. This was introduced in [117] and may be summarized as follows. Using the same notation as above, we define

$$\psi_{\lambda}^{\mathrm{dc}} := \psi_{\lambda}^{(\delta_{i_\lambda})} \delta_{i_\lambda}, \quad \lambda \in \mathcal{J}. \tag{2.51}$$

The induced spaces

$$S_j^{\rm dc} := \operatorname{clos}_{H(\operatorname{div};\Omega)} \operatorname{span} \Psi_j^{\rm dc}, \qquad S_j^{\rm df} := \operatorname{clos}_{H(\operatorname{div};\Omega)} \operatorname{span} \Psi_j^{\rm df}$$

fulfill

$$S_j^{\rm div} = S_j^{\rm dc} \oplus S_j^{\rm df},$$

where $S_j^{\rm div}$ is induced by $\Psi^{\rm div}$ in (2.43) in the natural way. This results in a splitting

$$H_0(\operatorname{div};\Omega) = V_0(\operatorname{div};\Omega) \oplus H^{\rm dc}(\Omega),$$

where $H^{\rm dc}(\Omega) = S(\Psi^{\rm dc})$ (and also for an analogous splitting of $L_2(\Omega)$). This decomposition is *not* orthogonal, but stable, which can be shown by introducing the dual functions

$$\tilde{\psi}_{\boldsymbol{\lambda}}^{\rm df} := \tilde{\psi}_\lambda^{(\delta_i)} \boldsymbol{\delta}_i, \quad \boldsymbol{\lambda} = (i,\lambda) \in \mathcal{J}^{\rm df}$$

as well as

$$(\tilde{\psi}_\lambda^{\rm dc})_{|\nu} := \partial_\nu[-(D_{i_\lambda}^{(i_\lambda)})^{-1}]_\lambda \tilde{\Psi}_{|\lambda|}^{(0)}, \quad 1 \le \nu \le n.$$

Then, the following orthogonality relations hold [117]

$$(\psi_{\boldsymbol{\lambda}}^{\rm df}, \tilde{\psi}_{\lambda'}^{\rm dc})_{0,\Omega} = (\tilde{\psi}_{\boldsymbol{\lambda}}^{\rm df}, \psi_{\lambda'}^{\rm dc})_{0,\Omega} = 0$$

for $\boldsymbol{\lambda} \in \mathcal{J}^{\rm df}$, $\lambda \in \mathcal{J}$, and

$$(\psi_\lambda^{\rm dc}, \tilde{\psi}_{\lambda'}^{\rm dc})_{0,\Omega} = \delta_{\lambda,\lambda'}, \quad (\psi_{\boldsymbol{\lambda}}^{\rm df}, \tilde{\psi}_{\boldsymbol{\lambda}'}^{\rm df})_{0,\Omega} = \delta_{\boldsymbol{\lambda},\boldsymbol{\lambda}'}$$

for all $\lambda, \lambda' \in \mathcal{J}$ and $\boldsymbol{\lambda}, \boldsymbol{\lambda}' \in \mathcal{J}^{\rm df}$.

2.5.2 Interrelations and Hodge Decompositions

With the various wavelet bases constructed above for flux and stream function spaces at hand, we obtain some relationships between the bases (and induced spaces, of course) as well as Hodge decompositions. Let us now consider the two Hodge decompositions (2.27) and (2.28).

Theorem 11. *Under Ass. 3 we have: For any subsets $\Lambda^{\rm df} \subset \mathcal{J}^{\rm df}$, $\Lambda^{\rm cf} \subset \mathcal{J}$, the induced spaces*

$$S_\Lambda^{\rm df} := \operatorname{clos}_{V_0(\operatorname{div};\Omega)} \operatorname{span} \Psi_{\Lambda^{\rm df}}^{\rm df}, \qquad S_\Lambda^{\rm cf} := \operatorname{clos}_{H(\mathbf{curl};\Omega)} \operatorname{span} \Psi_{\Lambda^{\rm cf}}^{\rm cf}, \tag{2.52}$$

are orthogonal with respect to $(\cdot,\cdot)_{0,\Omega}$ and with respect to

$$a_{\rm div}(\boldsymbol{u},\boldsymbol{v}) := (\boldsymbol{u},\boldsymbol{v})_{0,\Omega} + (\operatorname{div} \boldsymbol{u}, \operatorname{div} \boldsymbol{v})_{0,\Omega}, \tag{2.53}$$

as well as with respect to

$$a_{\rm curl}(\boldsymbol{u},\boldsymbol{v}) := (\boldsymbol{u},\boldsymbol{v})_{0,\Omega} + (\mathbf{curl}\, \boldsymbol{u}, \mathbf{curl}\, \boldsymbol{v})_{0,\Omega}. \tag{2.54}$$

Proof. Using integration by parts and (2.7), we obtain

$$\begin{aligned}(\boldsymbol{\psi}_{\boldsymbol{\lambda}}^{\mathrm{df}}, \boldsymbol{\psi}_{\lambda'}^{\mathrm{cf}})_{0,\Omega} &= (\boldsymbol{\psi}_{\boldsymbol{\lambda}}^{\mathrm{df}}, \mathbf{grad}\, \psi_{\lambda'}^{(1)})_{0,\Omega} \\ &= (-1)\,(\mathrm{div}\, \boldsymbol{\psi}_{\boldsymbol{\lambda}}^{\mathrm{df}}, \psi_{\lambda'}^{(1)})_{0,\Omega} \;=\; 0, \qquad \boldsymbol{\lambda} \in \boldsymbol{\mathcal{J}}^{\mathrm{df}},\, \lambda' \in \mathcal{J},\end{aligned}$$

since $\boldsymbol{\psi}_{\boldsymbol{\lambda}}^{\mathrm{df}}$ is divergence-free. The orthogonality w.r.t. $a_{\mathrm{div}}(\cdot,\cdot)$ and $a_{\mathrm{curl}}(\cdot,\cdot)$ automatically follows from the fact that $\boldsymbol{\psi}_{\boldsymbol{\lambda}}^{\mathrm{df}}$, $\boldsymbol{\psi}_{\lambda}^{\mathrm{cf}}$ are divergence-, respectively **curl**-free. □

The latter theorem obviously gives rise to a Hodge decomposition in the spirit of (2.27). Note that Theorem 11 holds for *any* subsets Λ^{df}, Λ^{cf}, so that these bases may also be used for an adaptive approach. Moreover, it should be noted that $\boldsymbol{\Psi}^{\mathrm{cf}}$ in the latter theorem need not have vanishing traces on the boundary of Ω. Recall that, due to Remark 6 this is also provided by our construction. The second Hodge decomposition (2.28) now involves the space $\boldsymbol{V}_0(\mathbf{curl};\Omega)$, i.e., the trace of these functions vanishes on the boundary of Ω. The price we have to pay for this is a higher smoothness of the divergence-free wavelet functions (and hence, in general, a larger support).

Theorem 12. *In addition to Ass. 3 assume that* $\Psi^{(0)} \subset \boldsymbol{H}_0^1(\Omega)$. *Then, the subsets* $\boldsymbol{S}_\Lambda^{\mathrm{df}}$ *and* $\boldsymbol{S}_\Lambda^{\mathrm{cf}}$ *give rise to a Hodge decomposition in the sense of* (2.28) *w.r.t.* $(\cdot,\cdot)_{\boldsymbol{L}_2(\Omega)}$, $a_{\mathrm{div}}(\cdot,\cdot)$ *and* $a_{\mathrm{curl}}(\cdot,\cdot)$.

Proof. Since $\Psi^{(0)} \subset \boldsymbol{H}_0^1(\Omega)$, we obtain

$$\boldsymbol{S}_\Lambda^{\mathrm{df}} \subset \boldsymbol{H}^1(\Omega) \cap \boldsymbol{V}(\mathrm{div};\Omega) \cap \boldsymbol{H}_0(\mathbf{curl};\Omega)$$

which is isomorphic to $\boldsymbol{V}_0(\mathbf{curl};\Omega)^{\perp}$, see (2.29). The orthogonality was already shown in the proof of Theorem 11 above. □

2.6 General Domains

Our construction is so far based on the general hypotheses stated in Ass. 3. In this section, we give examples of wavelet bases on certain domains $\Omega \subset \mathbb{R}^n$ that satisfy this assumption. Moreover, we consider an alternative using a fictitious domain approach.

2.6.1 Tensor Product Domains

The first example is related to tensor product domains. On these domains, wavelet bases can be used that are tensor products of corresponding functions on the interval $[0,1]$. Hence, we have to investigate Ass. 3 for wavelet systems on the interval.

In the Sect. s 2.1.1 on p. 84 and 2.1.2 on p. 85, we have shown differentiation and integration relations between scaling functions and wavelets. In fact, Theorem 7 on p. 85 ensures that Ass. 3 holds for tensor product bases.

2.6.2 Parametric Mappings

Of course, tensor product domains are of limited use when one thinks to applications in numerical analysis. Hence, we consider two types of generalizations. Both treat domains that are the image of $\hat{\Omega} = [0,1]^n$ under certain parametric mappings. The first example are affine mappings, the second one conformal mappings in 2D. Finally, we outline further extensions using domain decomposition ideas.

In this section, we consider domains $\Omega \subset \mathbb{R}^n$ that are images under certain parametric mappings $F : \mathbb{R}^n \to \mathbb{R}^n$, i.e., $\Omega = F(\hat{\Omega})$, where $\hat{\Omega} := (0,1)^n$ denotes the reference cube. Of course, as in Sect. 1.5 on p. 47, F has to be bijective.

Affine Images. If Ω is a parallelepiped in $\mathbb{R}^n$, it is known that there exists an affine mapping F such that $\Omega = F(\hat{\Omega})$. In particular, there exists some $A \in GL(\mathbb{R}^n, \mathbb{R}^n)$ and some $b \in \mathbb{R}^n$ such that $F(\hat{x}) = A\hat{x} + b$, $\hat{x} \in \hat{\Omega}$. Hence, the linear tangent mapping DF and hence also its determinant $|DF|$ are constant with respect to $\hat{x}$.

Now, define for any $\hat{\boldsymbol{v}} \in \boldsymbol{H}(\mathrm{div}; \hat{\Omega})$

$$\boldsymbol{v}(x) := \frac{1}{|DF|} DF\, \hat{\boldsymbol{v}}(\hat{x}), \quad x = F(\hat{x}) \in \Omega. \tag{2.55}$$

It is well known (see for example [111]) that the mapping $\hat{\boldsymbol{v}} \mapsto \boldsymbol{v}$ is a bijective affine transformation from $\boldsymbol{H}(\mathrm{div}; \hat{\Omega})$ onto $\boldsymbol{H}(\mathrm{div}; \Omega)$. Moreover, one has

$$\int_{\Omega} q(x) \,\mathrm{div}\, \boldsymbol{v}(x)\, dx = \int_{\hat{\Omega}} \hat{q}(\hat{x}) \,\mathrm{div}\, \hat{\boldsymbol{v}}(\hat{x})\, d\hat{x} \tag{2.56}$$

for all $\hat{q} \in L_2(\hat{\Omega})$, where

$$q(x) := \hat{q}(\hat{x}), \quad x = F(\hat{x}) \in \Omega. \tag{2.57}$$

Let us now consider wavelet bases $\hat{\boldsymbol{\Psi}}^{\mathrm{div}}$ and $\hat{\boldsymbol{\Psi}}^{\mathrm{df}}$ on $\hat{\Omega}$ as defined in (2.43) and Definition 5 above. Then, we define $\boldsymbol{\Psi}^{\mathrm{div}}$ and $\boldsymbol{\Psi}^{\mathrm{df}}$ by using (2.55). Since $\hat{\boldsymbol{v}} \mapsto \boldsymbol{v}$ is a mapping onto and because of (2.56), the latter wavelet systems are in fact wavelet bases for $\boldsymbol{H}_0(\mathrm{div}; \Omega)$ and $\boldsymbol{V}_0(\mathrm{div}; \Omega)$, respectively. Finally, concerning $\boldsymbol{V}_0(\mathbf{curl}; \Omega)$, we remark that (2.56) implies

$$\int_{\Omega} \mathbf{grad}\, q(x) \cdot \boldsymbol{v}(x)\, dx = \int_{\hat{\Omega}} \mathbf{grad}\, \hat{q}(\hat{x}) \cdot \hat{\boldsymbol{v}}(\hat{x})\, d\hat{x}. \tag{2.58}$$

Now, define $\Psi^{(1)}$ starting from $\hat{\Psi}^{(1)}$ using (2.57). Then, we obtain that $\mathbf{grad}\, \psi_\lambda^{(1)}$ is orthogonal to $\boldsymbol{\Psi}^{\mathrm{df}}$ for all $\lambda \in \mathcal{J}$ in view of (2.58). Hence, (2.29) implies that $\mathbf{grad}\, \Psi^{(1)}$ is in fact a wavelet basis for $\boldsymbol{V}_0(\mathbf{curl}; \Omega)$. Note that the duals are defined by

$$\tilde{\boldsymbol{v}}(x) := DF^{-T}\, \hat{\tilde{\boldsymbol{v}}}(\hat{x}), \quad x = F(\hat{x}) \in \Omega.$$

Conformal Images. Conformal mappings have been extensively used both for the analysis and the numerical solution of incompressible flow problems. This is due to the fact that mappings that are conformal (and orthogonal) preserve certain simple differential operators as div and **grad**, see [20, 75]. On the other hand, these mappings are restricted to the 2D–case.

In contrast to (2.55), we define in this case for any $\hat{\boldsymbol{v}} \in \boldsymbol{H}(\text{div}; \hat{\Omega})$

$$\boldsymbol{v}(x) := \hat{\boldsymbol{v}}(\hat{x}), \qquad x = F(\hat{x}) \in \Omega.$$

Since div and **grad** are preserved, it is obvious that this gives wavelet systems in $\boldsymbol{H}(\text{div}; \Omega)$ and $\boldsymbol{H}(\textbf{curl}; \Omega)$ as well as divergence-, respectively curl-free wavelet bases (in 2D using (2.16)). In this case, one may define the dual system by

$$\tilde{\boldsymbol{v}}(x) := \frac{1}{|DF(\hat{x})|}\hat{\tilde{\boldsymbol{v}}}(\hat{x}), \qquad x = F(\hat{x}) \in \Omega.$$

More General Domains. Next, we could try to use the constructions of wavelet bases on general domains based on domain decomposition ideas as described in Sect. 1.5 on p. 47 within the framework presented in this section in order to obtain wavelets in $\boldsymbol{H}(\text{div}; \Omega)$ and $\boldsymbol{H}(\textbf{curl}; \Omega)$ on more complex domains. However, since we used integration by parts, the traces of the divergence- and **curl**-free wavelets vanish on the boundary. Hence, one has to add certain functions near the boundary in order to generate the full spaces $\boldsymbol{V}(\text{div}; \Omega)$ and $\boldsymbol{V}(\textbf{curl}; \Omega)$, respectively. These functions then have to be combined across the interelement boundaries while preserving their div and **curl**, respectively, which might be somewhat unsatisfactory as well as very technical.

A different approach was proposed in [19]. Here, the global continuity is enforced by means of Lagrange multipliers and hence one does not need to take care explicitly on the interelement boundaries. However, on each of the subdomains Ω_i one still has to have a *full* basis, i.e., one in $\boldsymbol{V}(\text{div}; \Omega_i)$ and $\boldsymbol{V}(\textbf{curl}; \Omega_i)$ rather than in $\boldsymbol{V}_0(\text{div}; \Omega_i)$ and $\boldsymbol{V}_0(\textbf{curl}; \Omega_i)$. Hence, one is still left with the problem of adding appropriate functions near the boundary.

2.6.3 Fictitious Domain Method

The latter remarks are not quite promising. However, we would like to remark that one can use divergence- and curl-free wavelet bases also within a fictitious domain method on more general domains. The well-known strategy in a fictitious domain approach for a Wavelet-Galerkin method on a general domain Ω is to embed the domain in a somewhat larger but simpler domain $\Omega \subset \square$, e.g., a box. Then, the boundary conditions on Ω are enforced weakly by Lagrange multipliers. In the wavelet context this has e.g. been studied in [51, 92]. It is an almost trivial remark that $\boldsymbol{\Psi}^{\text{cf}}$ and $\boldsymbol{\Psi}^{\text{df}}$ on $\square$ can be used in such a framework. Assume that $\boldsymbol{v} \in \boldsymbol{V}(D; \square)$ where $D \in \{\text{div}, \textbf{curl}\}$, i.e.,

$$(Dv, w)_{0,\square} = 0 \qquad \text{for all } w \in L_2(\square).$$

Let us consider and $\tilde{w} \in L_2(\Omega)$ and define $w := \tilde{w}\chi_\Omega$ on all $\square$. Then

$$(Dv_{|\Omega}, \tilde{w})_{0,\Omega} = (Dv, \tilde{w})_{0,\square} = 0,$$

i.e., $v_{|\Omega} \in V(D;\Omega)$. This allows to work with a tensor product wavelet basis on $\square$ which can be constructed following the lines above.

2.7 Examples

In this section, we show concrete examples for our construction. We would like to mention, that all computations that were needed to produce the figures below have been made using the *Multilevel Library* that is documented in [6].

Obviously, Theorem 7 implies the validity of Ass. 3 in 1D and $\Omega \in \{(0,1), \mathbb{R}, \mathbb{T}\}$. Using again the abbreviations $\lambda = (j,k)$ and $[D_j]^\lambda$, $[D_j]_\lambda$ for the λ-th row and column of D_j, respectively, we can also reformulate (2.4) on p. 86 for each wavelet function:

$$\begin{aligned} \tfrac{d}{dx}\xi_\lambda^{(1)} &= [D_{j,0}]^\lambda\, \Xi_j^{(0)}, & \tilde{\xi}_\lambda^{(1)} &= -[D_{j,0}]_\lambda^T\, \tfrac{d}{dx}\tilde{\Xi}_j^{(0)}, \\ \tfrac{d}{dx}\eta_\lambda^{(1)} &= [D_{j,1}]^\lambda\, \Upsilon_j^{(0)}, & \tilde{\eta}_\lambda^{(1)} &= -[D_{j,1}]_\lambda^T\, \tfrac{d}{dx}\tilde{\Upsilon}_j^{(0)}. \end{aligned} \tag{2.59}$$

Now, we use the systems induced by $\Xi^{(1)}$, $\tilde{\Xi}^{(1)}$ as well as $\Xi^{(0)}$, $\tilde{\Xi}^{(0)}$ within a tensor product framework. For $\gamma = (\gamma_1, \dots, \gamma_n)^T \in \{0,1\}^n =: E^n$, we define

$$\psi_\lambda^{(\gamma)}(x) := \psi_{(j,e,k)}^{(\gamma)}(x) := \prod_{\nu=1}^n \vartheta_{j,e_\nu,k_\nu}^{\gamma_\nu}(x_\nu),\ \vartheta_{j,e_\nu,k_\nu}^{\gamma_\nu} := \begin{cases} \eta_{j,k_\nu}^{(1)}, & \gamma_\nu = 1,\ e_\nu = 1, \\ \eta_{j,k_\nu}^{(0)}, & \gamma_\nu = 0,\ e_\nu = 1, \\ \xi_{j,k_\nu}^{(1)}, & \gamma_\nu = 1,\ e_\nu = 0, \\ \xi_{j,k_\nu}^{(0)}, & \gamma_\nu = 0,\ e_\nu = 0. \end{cases} \tag{2.60}$$

All systems of functions $\Psi_j^{(\gamma)}$ as well as their duals are defined in a straightforward manner. Now, we may collect all what has been said in this subsection to obtain:

Proposition 6. *The biorthogonal systems* $\Psi^{(\gamma)}$, $\tilde{\Psi}^{(\gamma)}$, $\gamma \in E^n$ *fulfill Ass. 3 for* $\Omega = (0,1)^n$. □

Orthogonal Wavelets. One may wish to have an orthonormal wavelet basis for $\boldsymbol{H}(\mathrm{div};\Omega)$ and $\boldsymbol{H}(\mathbf{curl};\Omega)$. As already said, orthonormal wavelets are a special case of biorthogonal ones where primal and dual functions coincide. Hence one can, in principle, apply our construction. However, this does *not* give rise to an orthonormal basis on $\boldsymbol{H}(\mathrm{div};\Omega)$ or $\boldsymbol{H}(\mathbf{curl};\Omega)$. In fact, it was proven in [99] that there exists no orthonormal basis of compactly supported divergence-free wavelets.

Multiwavelets. Quite recently, divergence-free multiwavelets have been introduced, [93, 94, 95]. A multiwavelet basis is generated by a multiresolution analysis formed by more than one scaling function. This additional freedom allows e.g. the construction of multiwavelets whose support is smaller than the one of the corresponding wavelet basis functions with the same smoothness.

Having a closer look at the setup introduced by Ass. 3, it turns out that multiwavelet bases are covered by our setting. In fact, in [94, 95] it was shown that a specific piecewise linear multiwavelet basis in [85] fulfills Ass. 3. Hence, all the constructions in this section can in fact be realized for these multiwavelets.

It is an interesting result that also for multiwavelets it is not possible to construct an orthogonal divergence-free basis of compactly supported wavelets, [96]. This generalizes the previously mentioned results in [99].

Bivariate **curl**-*free Wavelets.* For graphical reasons we display only 2D functions here. Then, we end up with three types of wavelets (corresponding to $e \in E^2$):

$$\psi_1^{\mathrm{cf}}(x,y) = \begin{pmatrix} (\xi^{(0)}(x) - \xi^{(0)}(x-1))\eta^{(1)}(y) \\ 4\xi^{(1)}(x)\eta^{(0)}(y) \end{pmatrix},$$
$$\psi_2^{\mathrm{cf}}(x,y) = \begin{pmatrix} 4\eta^{(0)}(x)\xi^{(1)}(y) \\ \eta^{(1)}(x)(\xi^{(0)}(y) - \xi^{(0)}(y-1)) \end{pmatrix},$$
$$\psi_3^{\mathrm{cf}}(x,y) = \begin{pmatrix} 4\eta^{(0)}(x)\eta^{(1)}(y) \\ 4\eta^{(1)}(x)\eta^{(0)}(y) \end{pmatrix},$$

that are displayed for using biorthogonal B-spline wavelets with $d = 2$, $\tilde{d} = 8$ in Fig. 2.5.

Bivariate divergence-free Wavelets. Again, we obtain three different types of divergence-free wavelets (corresponding to $e \in E^2$):

$$\psi_1^{\mathrm{df}}(x,y) = \begin{pmatrix} -(\eta^{(1)}(x)(\xi^{(0)}(y) - \xi^{(0)}(y-1)) \\ 4\eta(0)(x)\xi^{(1)}(y) \end{pmatrix},$$
$$\psi_2^{\mathrm{df}}(x,y) = \begin{pmatrix} 4\xi^{(1)}(x)\eta^{(0)}(y) \\ -(\xi^{(0)}(x) - \xi^{(0)}(x-1))\eta^{(1)}(y) \end{pmatrix},$$
$$\psi_3^{\mathrm{df}}(x,y) = \begin{pmatrix} 4\eta^{(1)}(x)\eta^{(0)}(y) \\ -4\eta^{(0)}(x)\eta^{(1)}(y) \end{pmatrix}.$$

Comparing these formulas with the definition of the **curl**-free wavelets above, we see that all components again appear, just arranged in a different order (and possibly multiplied by a constant factor), i.e.,

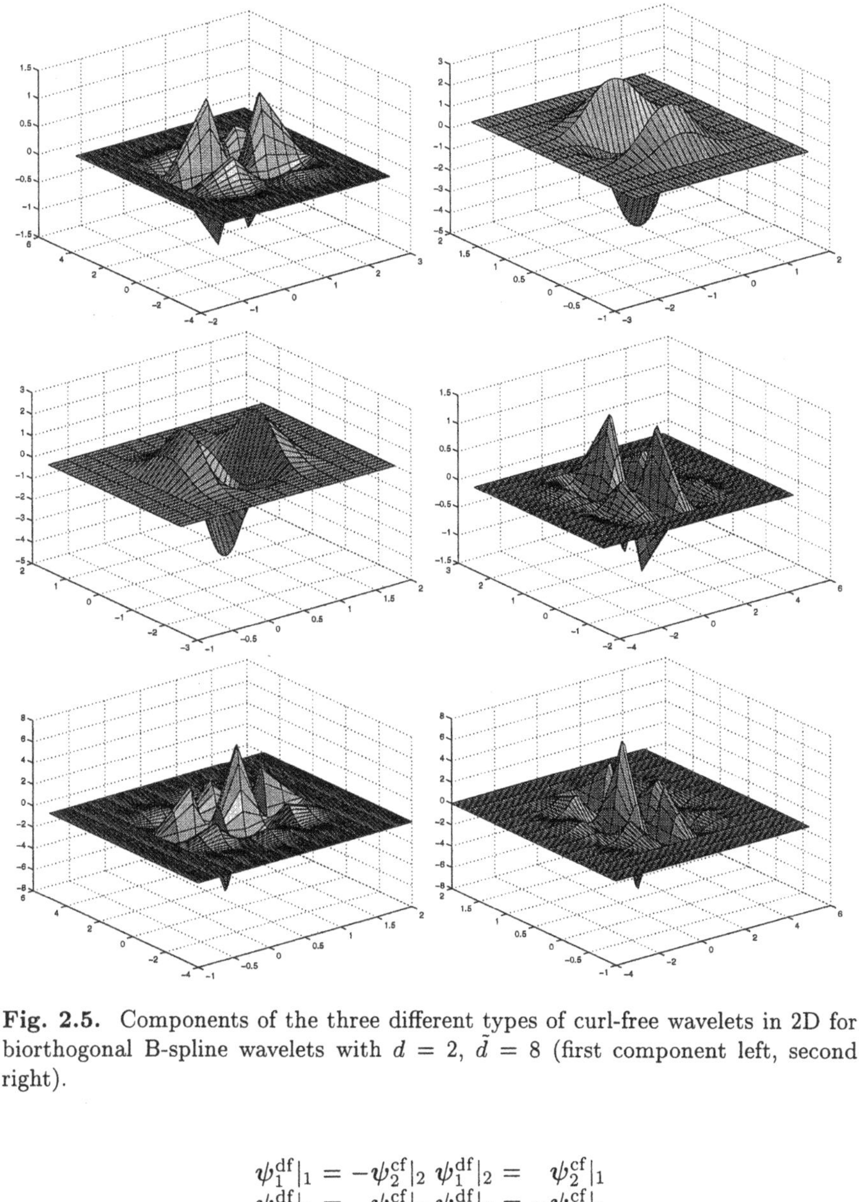

Fig. 2.5. Components of the three different types of curl-free wavelets in 2D for biorthogonal B-spline wavelets with $d = 2$, $\tilde{d} = 8$ (first component left, second right).

$$\begin{array}{ll} \psi_1^{\mathrm{df}}|_1 = -\psi_2^{\mathrm{cf}}|_2 & \psi_1^{\mathrm{df}}|_2 = \psi_2^{\mathrm{cf}}|_1 \\ \psi_2^{\mathrm{df}}|_1 = \psi_1^{\mathrm{cf}}|_2 & \psi_2^{\mathrm{df}}|_2 = -\psi_1^{\mathrm{cf}}|_1 \\ \psi_3^{\mathrm{df}}|_1 = \psi_3^{\mathrm{cf}}|_2 & \psi_3^{\mathrm{df}}|_2 = -\psi_3^{\mathrm{cf}}|_1 \end{array}$$

so that we refer again to Fig. 2.5 for corresponding pictures.

3 Applications

In this chapter, we describe some applications of the wavelet bases described above. These applications are of two types. First, we use the strong analytical properties combined with the particular construction of wavelets in $\boldsymbol{H}(\mathrm{div};\Omega)$ and $\boldsymbol{H}(\mathbf{curl};\Omega)$ in order to prove robustness and optimality of wavelet preconditioners in a Wavelet-Galerkin method for certain relevant problems. These problems include the Lamé equations in linear elasticity, Maxwell's equations in electromagnetism and formulations of the incompressible Navier-Stokes equations in the primitive variables.

The second type of examples is of more experimental character. Here, wavelets are used to extract physically relevant quantities from numerical simulations. In Sect. 3.2, divergence-free wavelets are used for the analysis and simulation of turbulent flow fields. In Sect. 3.3 the hardening of an elasto-plastic rod is considered. Here, wavelets are used to detect plastic waves in the material.

3.1 Robust and Optimal Preconditioning

In this section, we use the decomposition of $\boldsymbol{L}_2(\Omega)$ introduced in Sect. 2.2.3 (p. 92) in order to derive robust and optimal preconditioners for problems involving certain parameters. By *robust*, we mean here, that the condition number of the preconditioned stiffness matrix does not depend on the involved parameters. We consider three examples, namely preconditioning in the spaces $\boldsymbol{H}(\mathrm{div};\Omega)$ and $\boldsymbol{H}(\mathbf{curl};\Omega)$, the Lamé and Maxwell's equation.

3.1.1 Wavelet-Galerkin Discretizations

Before turning to our applications, let us briefly review the main facts on Wavelet-Galerkin methods that are relevant for the sequel.

Elliptic Problems. Let us consider a selfadjoint operator $A : H \to H'$, which is *H-elliptic*, i.e.,

$$a(v,w) := \langle Av, w\rangle \lesssim \|v\|_H \|w\|_H \quad \text{and} \quad a(v,v) \sim \|v\|_H^2 \qquad (3.1)$$

for all $u, v \in H$. Clearly (3.1) means that A is an isomorphism from H to H', i.e.,

$$\|Av\|_{H'} \sim \|v\|_H, \quad v \in H. \tag{3.2}$$

Thus the equation

$$Au = f \tag{3.3}$$

has a unique solution for any $f \in H'$ which will always be denoted by u. Typical examples are second order elliptic boundary value problems with homogeneous Dirichlet boundary conditions on some open domain $\Omega \subset \mathbb{R}^d$. In this case $H = H_0^1(\Omega)$ and $H' = H^{-1}(\Omega)$. Other examples are obtained by turning an exterior boundary value problem into a singular integral equation on the boundary Γ of the domain. For a formulation in terms of the single layer potential operator one obtains for instance $H = H^{-1/2}(\Gamma)$ and $H' = H^{1/2}(\Gamma)$, see [36, 104, 112] for details. Thus H is typically a Sobolev space and

$$H \hookrightarrow L_2 \hookrightarrow H' \quad \text{or} \quad H' \hookrightarrow L_2 \hookrightarrow H.$$

We sometimes write $H = H^t$ to indicate the Sobolev regularity although often a closed subspace of the full Sobolev space determined by boundary conditions is meant. H^{-t} is always the dual of this particular subspace. We hasten to add though that A need not be a scalar equation but could as well represent a system (such as the Stokes equations) in which case H is typically a product of Sobolev spaces as long as the above assumptions are fulfilled.

We are interested in solving (3.3) approximately with the aid of a Galerkin method, i.e., we pick some finite dimensional space $S \subset H$ and search for $u_S \in S$ such that

$$\langle Au_S, v_S \rangle = \langle f, v_S \rangle, \qquad v_S \in S, \tag{3.4}$$

where $\langle \cdot, \cdot \rangle$ denotes the dual pairing of H' and H. We will consider here spaces generated by adaptively chosen functions out of a wavelet basis, i.e., $S = S_\Lambda$, (1.32). This means we are seeking $u_\Lambda = \boldsymbol{d}_\Lambda^T \Psi_\Lambda$ such that

$$\langle Au_\Lambda, \psi_{\lambda'} \rangle = \langle f, \psi_{\lambda'} \rangle$$

for all $\lambda' \in \Lambda$. Now, we are in the setting of a *Wavelet-Galerkin Method.*

Wavelet Multilevel Preconditioning. Once a wavelet basis Ψ with the properties listed in Sect. 1.1.1 is given, the equivalences (3.2) and (1.30) (pp. 110 and 10) allow to transform the operator equation (3.3) over a function space H into a *matrix* equation over the corresponding *sequence space*. The matrix in question is the representation of the operator with respect to the chosen wavelet basis. More precisely, the relation (1.30) suggests a special *scaling* of the basis which leads to

$$\boldsymbol{A} := \boldsymbol{D}^{-T} \langle \Psi, A\Psi \rangle \boldsymbol{D}^{-1}. \tag{3.5}$$

The crucial point is that the norm equivalence (1.30) in conjunction with ellipticity (3.2) implies that the (infinite) matrix $\boldsymbol{A}$ defined by (3.5) is now an automorphism on $\ell_2(\mathcal{J})$, [47, 50, 57, 88].

Theorem 13. *Let $A : H \to H'$ be H-elliptic and suppose $\boldsymbol{D}^{-T}\Psi$ and $\boldsymbol{D}^{T}\tilde{\Psi}$ are generalized wavelet bases for H and H', in the sense of Definition 1 on* *p. 2**, i.e.,*

$$\|\boldsymbol{D}\boldsymbol{d}\|_{\ell_2} \sim \|\boldsymbol{d}^T\Psi\|_H, \qquad \|\boldsymbol{D}^{-T}\boldsymbol{d}\|_{\ell_2} \sim \|\boldsymbol{d}^T\tilde{\Psi}\|_{H'}, \tag{3.6}$$

for all $\boldsymbol{d} \in \ell_2$ and an invertible operator $\boldsymbol{D} \in GL(\ell_2, \ell_2)$. Then:

1. *The function $u = \boldsymbol{d}^T\Psi \in H$ solves the original operator equation (3.3) if and only if the sequence*
$$\boldsymbol{u} := \boldsymbol{D}\boldsymbol{d} \tag{3.7}$$
solves the matrix equation
$$\boldsymbol{A}\boldsymbol{u} = \mathbf{f}, \tag{3.8}$$
where $\mathbf{f} := \boldsymbol{D}^{-T}\langle\Psi, f\rangle$.
2. *For $\boldsymbol{A} = (a_{\lambda,\mu})_{\lambda,\mu\in\mathcal{J}}$, the operators*
$$\boldsymbol{A}_\Lambda := (a_{\lambda,\mu})_{\lambda,\mu\in\Lambda}, \quad \Lambda \subset \mathcal{J},$$
have uniformly bounded condition numbers
$$\mathrm{cond}_2(\boldsymbol{A}_\Lambda) \le \kappa, \quad \Lambda \subset \mathcal{J}. \tag{3.9}$$

Proof. Since this result will be of particular importance in the sequel, let us add the proof — which basically is taken from [47] — for convenience and completeness.

The first part is straightforward. As for the second part, we have for any $\boldsymbol{d} \in \ell_2$

$$\begin{aligned}
\|\boldsymbol{D}\,\boldsymbol{d}\|_{\ell_2}^2 \sim \|u\|_H^2 &\sim a(u,u) = \langle Au, u\rangle \\
&\sim \|Au\|_{H'}^2 = \|\langle Au, \Psi\rangle\,\tilde{\Psi}\|_{H'}^2 \\
&\sim \|\boldsymbol{D}^{-T}\langle Au, \Psi\rangle^T\|_{\ell_2}^2 = \|\boldsymbol{D}^{-T}\langle A\Psi, \Psi\rangle\,\boldsymbol{d}\|_{\ell_2}^2 \\
&= \|\boldsymbol{D}^{-T}\langle A\Psi, \Psi\rangle\,\boldsymbol{D}^{-1}\,\boldsymbol{D}\boldsymbol{d}\|_{\ell_2}^2 = \|\boldsymbol{A}\,\boldsymbol{D}\boldsymbol{d}\|_{\ell_2}^2,
\end{aligned}$$

which proves the theorem. □

Note that the above proof in fact shows a stronger property than stated in the second part of Theorem 13, namely $\|\boldsymbol{A}\| \sim \|\boldsymbol{A}^{-1}\| \sim 1$. This means that the original operator equation is equivalent to a boundedly invertible operator equation in ℓ^2. This in fact is a main requirement in the convergence and efficiency analysis of the adaptive wavelet method in [28, 29].

Let us add some comments on the general form of this result. To this end, consider the Helmholtz problem, [5, 27, 48]

$$-\varepsilon\,\Delta u + u = f \text{ in } \Omega, \quad \varepsilon \ll 1, \qquad u = 0 \text{ on } \Gamma = \partial\Omega. \tag{3.10}$$

If we consider the corresponding differential operator A as a mapping from $H_0^1(\Omega)$ to its dual $H^{-1}(\Omega)$, A is $H_0^1(\Omega)$-elliptic. However, the constants in (3.2) depend on ε so that $\text{cond}(A) = \mathcal{O}(\varepsilon^{-1})$, which of course is not satisfactory. In order to obtain a boundedly invertible ℓ_2-problem with ellipticity constant independent on ε (i.e., robust), we have to consider A on the energy space, i.e., the space of $H_0^1(\Omega)$-functions equipped with the norm

$$\|u\|_\varepsilon^2 := a_\varepsilon(u,u), \qquad a_\varepsilon(u,v) := \varepsilon(\text{grad}\, u, \text{grad}\, v)_{0,\Omega} + (u,v)_{0,\Omega}.$$

Denoting the induced Hilbert space by H_ε, one can in fact easily show that A is H_ε-elliptic independent on ε. In order to obtain the desired preconditioner, we still have to verify (3.6). Assuming that Ψ characterizes both $L_2(\Omega)$ and $H_0^1(\Omega)$ in the sense of (1.26) on p. 9 for $m = 0, 1$, we have

$$a_\varepsilon(u,u) = \varepsilon\|\text{grad}\, u\|_{0,\Omega}^2 + \|u\|_{0,\Omega}^2 \sim \sum_\lambda (\varepsilon 2^{2|\lambda|} + 1)\, d_\lambda^2,$$

for $u = \boldsymbol{d}^T\Psi$, so that we obtain $\boldsymbol{D} := \text{diag}(\varepsilon 2^{2|\lambda|} + 1)^{1/2}$. For numerical results concerning this problem we refer to [7, 32].

3.1.2 The Lamé Equations for Almost Incompressible Material

The theory of elasticity is concerned with the state of a body under the action of certain forces. In particular, one is interested in strains and stresses that are induced by the deformation of the body. Following [18], a standard assumption is that a reference configuration $\bar{\Omega}$ for the body is known. Here, $\bar{\Omega}$ is the closure of a bounded and open domain $\Omega \subset \mathbb{R}^3$, which typically is the domain occupied by the body in a stress free state. The actual state is determined by the mapping

$$\boldsymbol{\phi} : \bar{\Omega} \to \mathbb{R}^3,$$

i.e., $\boldsymbol{\phi}(x)$ is the coordinate of that point which corresponds to the point x in the reference configuration. One sets

$$\boldsymbol{\phi} = \mathbf{id} + \boldsymbol{u}$$

and $\boldsymbol{u}$ is called *displacement*. Then, $\nabla\boldsymbol{\phi}$ is called the *deformation gradient*. It turns out that the change of local length scales of the body is caused by the *strain tensor*

$$E := \frac{1}{2}(C - I), \qquad C := \nabla\boldsymbol{\phi}^T \nabla\boldsymbol{\phi},$$

i.e.,

$$E_{i,j} = \frac{1}{2}\left(\frac{\partial u_i}{\partial x_j} + \frac{\partial u_j}{\partial x_i}\right) + \frac{1}{2}\sum_k \frac{\partial u_i}{\partial x_k}\frac{\partial u_j}{\partial x_k}.$$

In linear elasticity, the quadratic terms are neglected and the *symmetric derivative*

$$\varepsilon_{ij} := \frac{1}{2}\Big(\frac{\partial u_i}{\partial x_j} + \frac{\partial u_j}{\partial x_i}\Big) \tag{3.11}$$

is used as an approximation of the entries of E. The equations under consideration then read

$$-2\mu \operatorname{div} \varepsilon(\boldsymbol{u}) - \nu \operatorname{\mathbf{grad}} \operatorname{div} \boldsymbol{u} = \boldsymbol{f}, \tag{3.12}$$

where again $\boldsymbol{u}$ is the unknown displacement of an isotropic elastic body. The constants ν and μ are known as *Lamé constants.* Finally, the *stress tensor* in linear elasticity takes the form

$$\sigma := \frac{E_e}{1+\zeta}\Big(\varepsilon + \frac{\zeta}{1-2\zeta}\operatorname{trace}(\varepsilon I)\Big),$$

where $E_e := \frac{\mu(3\nu+2\mu)}{\mu+\nu}$ denotes the *elastic modulus* and $\zeta := \frac{\nu}{2(\nu+\mu)}$ the *Poisson number.* Probl. (3.12) accompanied with boundary conditions

$$\boldsymbol{u} = 0 \text{ on } \Gamma_{\mathrm{Dir}}, \qquad \sigma(\boldsymbol{u}) \cdot \boldsymbol{n} = \boldsymbol{g} \text{ on } \Gamma_{\mathrm{Neu}},$$

$\partial\Omega = \Gamma = \Gamma_{\mathrm{Dir}} \cup \Gamma_{\mathrm{Neu}}$ is a classical partial differential equation in linear elasticity.

One case of particular interest is *almost incompressible material.* In this case, the Lamé constants are related by

$$\nu \gg \mu,$$

which causes serious problems for the construction of robust and efficient numerical solution methods. This means that the efficiency of the solution method should not depend on the Lamé constants, in particular not on the ratio $\frac{\nu}{\mu}$. This means, one aims at constructing a *robust* preconditioner to avoid this dependency. We will use the wavelet based Hodge decompositions in order to construct an asymptotically optimal and robust preconditioner for the Lamé equations.

The variational formulation of (3.12) is given by the bilinear form

$$\begin{aligned} a_{\nu,\mu}(\boldsymbol{u},\boldsymbol{v}) &:= \nu\,(\operatorname{div}\boldsymbol{u},\operatorname{div}\boldsymbol{v})_{0,\Omega} + 2\mu\,(\varepsilon(\boldsymbol{u}),\varepsilon(\boldsymbol{v}))_{0,\Omega} \\ &= (\nu+\mu)\,(\operatorname{div}\boldsymbol{u},\operatorname{div}\boldsymbol{v})_{0,\Omega} + \mu\,(\operatorname{\mathbf{grad}}\boldsymbol{u},\operatorname{\mathbf{grad}}\boldsymbol{v})_{0,\Omega}. \end{aligned} \tag{3.13}$$

Due to the well-known relation

$$(\operatorname{\mathbf{grad}}\boldsymbol{u},\operatorname{\mathbf{grad}}\boldsymbol{v})_{0,\Omega} = (\operatorname{div}\boldsymbol{u},\operatorname{div}\boldsymbol{v})_{0,\Omega} + (\operatorname{\mathbf{curl}}\boldsymbol{u},\operatorname{\mathbf{curl}}\boldsymbol{v})_{0,\Omega},\ \boldsymbol{u},\boldsymbol{v} \in \boldsymbol{H}_0^1(\Omega), \tag{3.14}$$

the form in (3.13) can also be written as

$$
\begin{aligned}
a(\boldsymbol{u},\boldsymbol{v}) &= (\nu+\mu)\,(\operatorname{div}\boldsymbol{u},\operatorname{div}\boldsymbol{v})_{0,\Omega} + \mu\,(\mathbf{grad}\,\boldsymbol{u},\mathbf{grad}\,\boldsymbol{v})_{0,\Omega}\\
&= (\nu+2\mu)\,(\operatorname{div}\boldsymbol{u},\operatorname{div}\boldsymbol{v})_{0,\Omega} + \mu\,(\mathbf{curl}\,\boldsymbol{u},\mathbf{curl}\,\boldsymbol{v})_{0,\Omega}\\
&= (\nu+2\mu)\,(\mathbf{grad}\,\boldsymbol{u},\mathbf{grad}\,\boldsymbol{v})_{0,\Omega} - (\nu+\mu)\,(\mathbf{curl}\,\boldsymbol{u},\mathbf{curl}\,\boldsymbol{v})_{0,\Omega}
\end{aligned}
$$

for $\boldsymbol{u},\boldsymbol{v}\in\boldsymbol{H}_0^1(\Omega)$. Of course the above equalities do not hold any longer when the traces of $\boldsymbol{u}$ and $\boldsymbol{v}$ on $\partial\Omega$ do not vanish. On the other hand, the bilinear form $a(\boldsymbol{u},\boldsymbol{v})$ appears in different variants in different applications. Hence, we consider the following bilinear forms

$$
\begin{aligned}
a_1(\boldsymbol{u},\boldsymbol{v}) &:= \varepsilon(\mathbf{grad}\,\boldsymbol{u},\mathbf{grad}\,\boldsymbol{v})_{0,\Omega} + (\operatorname{div}\boldsymbol{u},\operatorname{div}\boldsymbol{v})_{0,\Omega}, && \varepsilon\ll 1, && (3.15)\\
a_2(\boldsymbol{u},\boldsymbol{v}) &:= \varepsilon(\mathbf{grad}\,\boldsymbol{u},\mathbf{grad}\,\boldsymbol{v})_{0,\Omega} + (\mathbf{curl}\,\boldsymbol{u},\mathbf{curl}\,\boldsymbol{v})_{0,\Omega}, && \varepsilon\ll 1, && (3.16)\\
a_3(\boldsymbol{u},\boldsymbol{v}) &:= \nu(\operatorname{div}\boldsymbol{u},\operatorname{div}\boldsymbol{v})_{0,\Omega} + \mu(\mathbf{curl}\,\boldsymbol{u},\mathbf{curl}\,\boldsymbol{v})_{0,\Omega}, && \nu,\mu\geq 0 && (3.17)
\end{aligned}
$$

separately. It turns out that for each particular form it is advantageous to choose a different wavelet basis.

grad-div-form (3.15). It can easily be seen that $a_1(\cdot,\cdot)$ is a positive definite bilinear form on $\boldsymbol{H}_0^1(\Omega)$. We consider the space of $\boldsymbol{H}_0^1(\Omega)$-functions under the norm

$$\|\boldsymbol{u}\|_H := a_1(\boldsymbol{u},\boldsymbol{u})^{1/2}.$$

One readily confirms that $a_1(\cdot,\cdot)$ is bounded with respect to $\|\cdot\|_H$, i.e., $a_1(\boldsymbol{u},\boldsymbol{v})\leq\|\boldsymbol{u}\|_H\,\|\boldsymbol{v}\|_H$ independent on ε. For the discretization, we use $\boldsymbol{\Psi}^1 := \boldsymbol{\Psi}^{\mathrm{div}}\subset\boldsymbol{H}_0^1(\Omega)$ (which means that sufficiently smooth wavelets have to be chosen for a conformal discretization) so that we obtain the norm equivalence for $\boldsymbol{u}=\boldsymbol{c}^T\boldsymbol{\Psi}^1$:

$$
\begin{aligned}
\|\boldsymbol{u}\|_H^2 = a_1(\boldsymbol{u},\boldsymbol{u}) &= \varepsilon\|\mathbf{grad}\,\boldsymbol{u}\|_{0,\Omega}^2 + \|\operatorname{div}\boldsymbol{u}\|_{0,\Omega}^2\\
&\sim \varepsilon\sum_\lambda 2^{2|\lambda|}|\mathsf{c}_\lambda|^2 + \|\mathsf{div}\,\mathbf{c}\|_{\ell_2}^2\\
&= \|\boldsymbol{D}_1\,\boldsymbol{c}\|_{\ell_2}^2, && (3.18)
\end{aligned}
$$

where $\boldsymbol{D}_1$ is defined by

$$\boldsymbol{c}^T\boldsymbol{D}_1^2\boldsymbol{c} := \sum_\lambda(\varepsilon\,2^{2|\lambda|}\mathsf{c}_\lambda^2 + (\mathsf{div}\,\mathbf{c})_\lambda^2),$$

i.e., $\boldsymbol{D}_1 = (\varepsilon\,\mathrm{diag}(2^{2|\lambda|}) + \mathsf{div}^*\mathsf{div})^{1/2}$. By duality, we obtain in the same manner as for showing (1.31) for $\boldsymbol{v}=\boldsymbol{d}^T\tilde{\boldsymbol{\Psi}}^1\in\boldsymbol{L}_2(\Omega)$

$$\|\boldsymbol{v}\|_{H'} = \sup_{\boldsymbol{v}\in H}\frac{(\boldsymbol{u},\boldsymbol{v})_{0,\Omega}}{\|\boldsymbol{u}\|_H} \sim \sup_{\boldsymbol{c}\in\ell_2}\frac{\boldsymbol{d}^T\boldsymbol{c}}{\|\boldsymbol{D}_1\boldsymbol{c}\|_{\ell_2}} = \sup_{\boldsymbol{c}\in\ell_2}\frac{\boldsymbol{d}^T\boldsymbol{D}_1^{-1}\boldsymbol{c}}{\|\boldsymbol{c}\|_{\ell_2}} = \|\boldsymbol{D}_1^{-T}\boldsymbol{d}\|_{\ell_2},$$

since $\boldsymbol{D}_1$ in view of (3.18) is invertible. Again, all these estimates are independent of ε.

grad-curl-form (3.16). In analogy to the above case we use $\boldsymbol{\Psi}^2 := \boldsymbol{\Psi}^{\text{curl}} \subset \boldsymbol{H}^1(\Omega)$ for the discretization. Here we obtain the norm equivalences for

$$\|\boldsymbol{u}\|_H := a_2(\boldsymbol{u}, \boldsymbol{u})^{1/2}$$

as

$$\|\boldsymbol{u}\|_H \sim \|\boldsymbol{D}_2\boldsymbol{c}\|_{\ell_2}, \qquad \|\boldsymbol{v}\|_{H'} \sim \|\boldsymbol{D}_2^{-T}\boldsymbol{d}\|_{\ell_2}$$

for $\boldsymbol{u} = \boldsymbol{c}^T\boldsymbol{\Psi}^2 \in \boldsymbol{H}^1(\Omega)$, $\boldsymbol{v} = \boldsymbol{d}^T\tilde{\boldsymbol{\Psi}}^2 \in H'$ and $\boldsymbol{D}_2 := (\varepsilon\,\text{diag}(2^{2|\lambda|}) + \mathsf{curl}^*\mathsf{curl})^{1/2}$, i.e., $\boldsymbol{c}^T\boldsymbol{D}_2^2\boldsymbol{c} := \sum_\lambda(\varepsilon 2^{2|\lambda|}c_\lambda^2 + (\mathsf{curl}\,\boldsymbol{c})_\lambda^2)$. Moreover, $a_2(\cdot,\cdot)$ is positive definite and bounded with respect to $\|\cdot\|_H$. Again all inequalities are independent of ε.

div-curl-form (3.17). This case is somewhat different from the above two, since the use of

$$\boldsymbol{\Psi}^3 := \begin{pmatrix} \boldsymbol{\Psi}^{\text{df}} \\ \boldsymbol{\Psi}^{\text{cf}} \end{pmatrix}$$

allows to decouple the equations. The Hodge decomposition is the main tool for this decoupling. To be precise, it is again easy to verify that $a_3(\cdot,\cdot)$ is positive definite. We consider the space of functions in $\boldsymbol{H}(\text{div};\Omega) \cap \boldsymbol{H}(\textbf{curl};\Omega)$ equipped with the norm

$$\|\boldsymbol{u}\|_H := a_3(\boldsymbol{u}, \boldsymbol{u})^{1/2}.$$

Then, we obtain for

$$\boldsymbol{u} = \boldsymbol{c}^T\boldsymbol{\Psi}^3 = \begin{pmatrix} \boldsymbol{c}_{\text{df}} \\ \boldsymbol{c}_{\text{cf}} \end{pmatrix}^T \begin{pmatrix} \boldsymbol{\Psi}^{\text{df}} \\ \boldsymbol{\Psi}^{\text{cf}} \end{pmatrix} = \boldsymbol{u}_{\text{df}} + \boldsymbol{u}_{\text{cf}}$$

the following norm equivalence

$$\begin{aligned} \|\boldsymbol{u}\|_H^2 = a_3(\boldsymbol{u}, \boldsymbol{u}) &= \nu\|\text{div}\,\boldsymbol{u}_{\text{cf}}\|_{0,\Omega}^2 + \mu\|\textbf{curl}\,\boldsymbol{u}_{\text{df}}\|_{0,\Omega}^2 \\ &\sim \nu\|\mathsf{div}\,\boldsymbol{c}_{\text{cf}}\|_{\ell_2}^2 + \mu\|\mathsf{curl}\,\boldsymbol{c}_{\text{df}}\|_{\ell_2}^2 \\ &= \left\|\begin{pmatrix} \nu^{1/2}\mathsf{div} & 0 \\ 0 & \mu^{1/2}\mathsf{curl} \end{pmatrix}\boldsymbol{c}\right\|_{\ell_2}^2 =: \|\boldsymbol{D}_3\,\boldsymbol{c}\|_{\ell_2}^2 \end{aligned}$$

and in the same manner as above we obtain for $\boldsymbol{v} = (\boldsymbol{d}_{\text{df}}^T, \boldsymbol{d}_{\text{cf}}^T)\tilde{\boldsymbol{\Psi}}^3 \in H'$ the estimate $\|\boldsymbol{v}\|_{H'} \sim \|\boldsymbol{D}_3^{-T}\boldsymbol{c}\|_{\ell_2}$ again all estimates being independent on ν and μ.

In summary, we have proven

Theorem 14. *Under the above conditions, we have for*

$$\boldsymbol{A}^i := \boldsymbol{D}_i^{-T}a_i(\boldsymbol{\Psi}^i, \boldsymbol{\Psi}^i)\boldsymbol{D}_i^{-1},$$

that

$$\text{cond}_2\,(\boldsymbol{A}_\Lambda^i) = \mathcal{O}(1)$$

for $\#\Lambda \to \infty$ independent of all involved constants. □

3.1.3 The Maxwell Equations

Many phenomena in electromagnetism are modeled by the Maxwell equations. For given functions ρ (the charge density) and σ (the electrical conductivity of the medium in some domain $\Omega \subset \mathbb{R}^3$) as well as the given vector field $\boldsymbol{J}_s$ (the suppressed current density), one is interested in he electric field $\boldsymbol{E}$, the electric displacement $\boldsymbol{D}$, the magnetic field $\boldsymbol{H}$ and the magnetic flux density $\boldsymbol{B}$, which are related by the following set of equations:

$$\begin{aligned} \operatorname{div} \boldsymbol{D} &= \rho, \\ \operatorname{div} \boldsymbol{B} &= 0, \\ \tfrac{\partial}{\partial t}\boldsymbol{B} + \mathbf{curl}\,\boldsymbol{E} &= 0, \\ \tfrac{\partial}{\partial t}\boldsymbol{D} - \mathbf{curl}\,\boldsymbol{H} + \sigma \boldsymbol{E} &= -\boldsymbol{J}_s. \end{aligned} \tag{3.19}$$

The system (3.19) treats a rather general case. However, in many applications, additional information on the interrelation of the physical quantities is given. For example, it is known that for dielectric materials or insulators that are linear and isotropic, one has

$$\boldsymbol{B} = \mu \boldsymbol{H}, \qquad \boldsymbol{D} = \varepsilon \boldsymbol{E},$$

where μ is the magnetic permitivity and ε the electric permeability. Then, (3.19) simplifies to

$$\frac{\partial}{\partial t}\begin{pmatrix} \boldsymbol{B} \\ \boldsymbol{D} \end{pmatrix} + \mathcal{A}\begin{pmatrix} \boldsymbol{B} \\ \boldsymbol{D} \end{pmatrix} = \begin{pmatrix} 0 \\ -\boldsymbol{J}_s \end{pmatrix}, \qquad \mathcal{A} := \begin{pmatrix} 0 & \mathbf{curl}\,\frac{1}{\varepsilon(\cdot)} \\ -\mathbf{curl}\,\frac{1}{\mu(\cdot)} & \frac{\sigma}{\varepsilon} \end{pmatrix}, \tag{3.20}$$

where $\mathcal{A}$ is also called *Maxwell operator*, [64]. Assuming that ε, σ and μ are constant in space, using a backward difference approximation for the derivatives in time and inserting the second equation in (3.20) into the first one, gives

$$\begin{aligned} &\boldsymbol{B}(t,\cdot) + \tfrac{(\Delta t)^2}{\mu(\varepsilon + \sigma\,\Delta t)}\,\mathbf{curl}\,\mathbf{curl}\,\boldsymbol{B}(t,\cdot) \\ &= \boldsymbol{B}(t-\Delta t,\cdot) + \tfrac{(\Delta t)^2}{\varepsilon+\sigma\,\Delta t}\,\mathbf{curl}\,\boldsymbol{J}_s(t,\cdot) - \tfrac{\Delta t}{\varepsilon+\sigma\,\Delta t}\,\mathbf{curl}\,\boldsymbol{D}(t-\Delta t,\cdot). \end{aligned} \tag{3.21}$$

This can be simplified to

$$\boldsymbol{u} + \nu\,\mathbf{curl}\,\mathbf{curl}\,\boldsymbol{u} = f. \tag{3.22}$$

Hence, we consider the variational problem which corresponds to the bilinear form

$$a(\boldsymbol{u}, \boldsymbol{v}) := (\boldsymbol{u}, \boldsymbol{v})_{0,\Omega} + \nu\,(\mathbf{curl}\,\boldsymbol{u}, \mathbf{curl}\,\boldsymbol{v})_{0,\Omega}. \tag{3.23}$$

The corresponding differential operator L is a mapping from $\boldsymbol{H}(\mathbf{curl};\Omega)$ to $\boldsymbol{H}(\mathbf{curl};\Omega)'$ which again is bounded (for appropriate boundary conditions, see [64]) but *not* independent of the parameter ν. This however is a crucial drawback since the parameter ν in (3.21) can be very large.

Theorem 15. *The operator* $\boldsymbol{D}_{\mathrm{curl}} := (\boldsymbol{I} + \nu\, \mathsf{curl}^*\mathsf{curl})^{1/2}$ *and the wavelet basis* $\boldsymbol{\Psi}^{\mathrm{curl}}$ *fulfill the assumptions of Theorem 13, i.e, for* $\boldsymbol{A} := a(\boldsymbol{\Psi}^{\mathrm{curl}}, \boldsymbol{\Psi}^{\mathrm{curl}})$

$$\mathrm{cond}_2(\boldsymbol{D}_{\mathrm{curl}}^{-T} \boldsymbol{A}_\Lambda \boldsymbol{D}_{\mathrm{curl}}^{-1}) = \mathcal{O}(1), \qquad \#\Lambda \to \infty,$$

independent of ν.

Proof. We equip $\boldsymbol{H}(\mathbf{curl}; \Omega)$ with the norm $\|\boldsymbol{u}\|_H := a(\boldsymbol{u}, \boldsymbol{u})^{1/2}$. The bilinear form $a(\cdot, \cdot)$ is bounded and positive definite. By Theorem 8 on p. 94, we have for $\boldsymbol{u} = \boldsymbol{d}^T \boldsymbol{\Psi}^{\mathrm{curl}}$

$$\begin{aligned} \|\boldsymbol{u}\|_H^2 &= a(\boldsymbol{u}, \boldsymbol{u}) = \|\boldsymbol{u}\|_{0,\Omega}^2 + \nu \|\mathbf{curl}\, \boldsymbol{u}\|_{0,\Omega}^2 \\ &\qquad \sim \|\boldsymbol{d}\|_{\ell_2}^2 + \nu \|\mathsf{curl}\, \boldsymbol{d}\|_{\ell_2}^2 \;=\; \|\boldsymbol{D}_{\mathrm{curl}}\, \boldsymbol{d}\|_{\ell_2}^2, \end{aligned}$$

as well as $\|\boldsymbol{v}\|_{H'} \sim \|\boldsymbol{D}_{\mathrm{curl}}^{-T} \boldsymbol{d}\|_{\ell_2}$ for $\boldsymbol{v} \in \boldsymbol{c}^T \tilde{\boldsymbol{\Psi}}^{\mathrm{curl}} \in \boldsymbol{H}(\mathbf{curl}; \Omega)'$. □

Let us mention that the bilinear form (3.23) also occurs in the 2D *stream function-vorticity-pressure* and the 3D *vector potential-vorticity* formulation of the incompressible Navier-Stokes equations, [77].

3.1.4 Preconditioning in $\boldsymbol{H}(\mathrm{div}; \Omega)$

In analogy to (3.22), we consider the problem

$$\boldsymbol{u} - \nu\, \mathbf{grad}\, \mathrm{div}\, \boldsymbol{u} = f, \tag{3.24}$$

which corresponds to a differential operator from $\boldsymbol{H}(\mathrm{div}; \Omega)$ to $\boldsymbol{H}(\mathrm{div}; \Omega)'$ and the bilinear form

$$a(\boldsymbol{u}, \boldsymbol{v}) := (\boldsymbol{u}, \boldsymbol{v})_{0,\Omega} + \nu\, (\mathrm{div}\, \boldsymbol{u}, \mathrm{div}\, \boldsymbol{v})_{0,\Omega}. \tag{3.25}$$

In a similar fashion as in the latter subsection, we obtain a robust preconditioner.

Theorem 16. *The operator* $\boldsymbol{D}_{\mathrm{div}} := (\boldsymbol{I} + \nu\, \mathsf{div}^*\mathsf{div})^{1/2}$ *and the wavelet basis* $\boldsymbol{\Psi}^{\mathrm{div}}$ *fulfill the assumptions of Theorem 13, i.e, for* $\boldsymbol{A} := a(\boldsymbol{\Psi}^{\mathrm{div}}, \boldsymbol{\Psi}^{\mathrm{div}})$

$$\mathrm{cond}_2(\boldsymbol{D}_{\mathrm{div}}^{-T} \boldsymbol{A}_\Lambda \boldsymbol{D}_{\mathrm{div}}^{-1}) = \mathcal{O}(1), \qquad \#\Lambda \to \infty,$$

independent of ν.

Proof. We equip $\boldsymbol{H}(\mathrm{div}; \Omega)$ with $\|\boldsymbol{u}\|_H := a(\boldsymbol{u}, \boldsymbol{u})^{1/2}$. By (2.46), we have for $\boldsymbol{u} = \boldsymbol{d}^T \boldsymbol{\Psi}^{\mathrm{div}}$

$$\begin{aligned} \|\boldsymbol{u}\|_H^2 &= a(\boldsymbol{u}, \boldsymbol{u}) = \|\boldsymbol{u}\|_{0,\Omega}^2 + \nu \|\mathrm{div}\, \boldsymbol{u}\|_{0,\Omega}^2 \\ &\qquad \sim \|\boldsymbol{d}\|_{\ell_2}^2 + \nu \|\mathsf{div}\, \boldsymbol{d}\|_{\ell_2}^2 \;=\; \|\boldsymbol{D}_{\mathrm{div}} \boldsymbol{d}\|_{\ell_2}^2 \end{aligned}$$

and $\|\boldsymbol{v}\|_{H'} \sim \|\boldsymbol{D}_{\mathrm{div}}^{-T} \boldsymbol{d}\|_{\ell_2}$ for $\boldsymbol{v} = \boldsymbol{c}^T \tilde{\boldsymbol{\Psi}}^{\mathrm{div}} \in \boldsymbol{H}(\mathrm{div}; \Omega)'$. □

3.2 Analysis and Simulation of Turbulent Flows

A primary objective of this section (which is based upon [1]) is to demonstrate the usefulness of wavelet concepts and their key mechanisms for the analysis and compression of turbulent flows.

3.2.1 Numerical Simulation of Turbulence

Analysis and simulation of turbulence belong to the most difficult problems in fluid dynamics. Here, we concentrate on turbulent flows in *viscous*, incompressible fluids that are described by the *incompressible Navier-Stokes equations* in some domain $\Omega \subset \mathbb{R}^n$. For given exterior force $\boldsymbol{f} \in \boldsymbol{H}^{-1}(\Omega)$ one seeks in a given time interval $[0,T]$, $T>0$, for the velocity field $\boldsymbol{u} : \Omega \times [0,T] \to \mathbb{R}^n$ and the pressure $p : \Omega \times [0,T] \to \mathbb{R}$ such that

$$\frac{\partial}{\partial t}\boldsymbol{u} + (\boldsymbol{u} \cdot \mathbf{grad})\boldsymbol{u} - \frac{1}{\mathrm{Re}}\Delta \boldsymbol{u} + \mathbf{grad}\, p = \boldsymbol{f}, \qquad (3.26)$$

$$\operatorname{div} \boldsymbol{u} = 0, \qquad (3.27)$$

where Re denotes the *Reynolds number.* The equations (3.26) and (3.27) have to be equipped by initial and boundary conditions.

It is well-known that the behavior of the flow changes dramatically in dependence of the size of the Reynolds number. While for small values of the Reynolds number, the flow is *laminar*, in particular *stationary*, *periodic* or *quasi-periodic*, one observes for large Reynolds numbers a completely irregular and chaotic behavior of the flow. One may define a flow as *turbulent*, if the Fourier spectrum of the flow is continuous.

As opposed to laminar flows the following characteristic properties are typical for turbulent flows:

- Turbulent flows are instationary and in general 3D phenomena;
- The nonlinearity in the Navier-Stokes equations causes that small errors may become large;
- Turbulent flows are highly *diffusive*, i.e., the velocity of impulse, mass and heat transfer are quite high;
- Turbulent flows are characterized by the occurance of vortices or wiggles, whose size may vary over a large scale.

Extended descriptions of the physical phenomena in turbulent flows can be found e.g. in [80, 115].

A straightforward approach for the numerical simulation of turbulence is to discretize the Navier-Stokes equation with very high resolution (say, on a very fine grid) and try to resolve the small vortices on this fine grid. This approach is known as *direct numerical simulation (DNS).* However, the scope of applicability of a DNS is limited. In fact, the famous $k^{-5/3}$-law of Kolmogorov [90] implies that the size of the smallest vortices in a turbulent flow is proportional to $\nu^{3/4}$ (where k is the *turbulent kinetic energy* and ν is the *viscosity*). For a resolution on a grid one would need in 3D $\mathcal{O}(\nu^{-9/4})$ grid points for a fine discretization. However, typical values of ν in applications in aerodynamics are in the range of 10^{-6} or even smaller leading to 10^{13} grid points. This is far beyond memory capacities of modern computers and also of those in the near future.

Hence, one is forced to use a modeling in order to simulate turbulence. In many cases, one is only interested in mean values or certain filtered quantities of the primitive variables velocity and pressure. Applying a filter to the Navier-Stokes equations gives rise to the so-called *Reynolds equations.* This system, however, is no longer closed, the *closing problem* occurs. This is the point where modeling comes into play. Without going into the details let us mention the *Smagorinsky model*, [114] and statistical models such as the *k-ε-model*, [97], or *Large-Eddy-Simulations.* Here we will consider the *Proper Orthogonal Decomposition (POD)*, see Sect. 3.2.3 below.

3.2.2 Divergence-Free Wavelet Analysis of Turbulence

In [12, 71, 72, 73, 81, 91, 100, 101, 105], wavelets have already been used for the analysis and simulation of turbulent flows. In most of the cited papers, a data analysis has been performed by means of orthogonal wavelets for the vorticity of flows with periodic boundary data. For the simulation of periodic flows so far mainly exponentially decaying *vaguelettes* have been employed e.g. in [73]. In [81], interpolatory Deslaurier-Debuc functions have been used for an adaptive simulation.

By contrast, in this section we propose to utilize compactly supported divergence-free wavelets as introduced in Definition 5 (p. 99) for the analysis of velocity fields. Since we can therefore work with the velocity/pressure formulation of the Navier-Stokes equations these tools work naturally also for three-dimensional problems on bounded domains so that the influence of no-slip boundary conditions can be taken into account. Since the incompressibility constraint is satisfied by the basis functions, such systems are natural candidates for a quantitative analysis of the role of incompressibility and its perturbation. Specifically, we will be concerned with the following issues.

First, divergence-free wavelets $\boldsymbol{\Psi}^{\text{div}}$ and appropriate complement wavelets $\boldsymbol{\Psi}^{\text{dc}}$ as introduced in Sect. 2.5.1 on p. 100 give rise to a Helmholtz-type decomposition for spaces of vector fields. Using the divergence-free wavelet transform, we can split flow data obtained from experiments as well as direct numerical simulations into its solenoidal and compressible parts. In particular, this tool offers us insight into the *Proper Orthogonal Decomposition* (POD) method [102], which is a well-known technique for constructing a low-dimensional turbulence model. POD modes are given as the eigenfunctions of the autocorrelation tensor for an ensemble of flow realizations. These modes can be interpreted as the *coherent structures* in a turbulent flow. In order to produce incompressible flow model equations based on the POD-Galerkin method, the corresponding data fields have to be divergence-free. However, it turns out that all the considered flows do in fact contain some compressible components. By comparing the POD modes obtained from the available data and those from its divergence-free part we study the quantitative influence of this violation on the incompressible flow model.

Second, the above mentioned ability of wavelet expansions to separate contributions from different characteristic length scales will be used to provide insight into the multiscale structure of the various flows and, in particular, of the coherent structures of the wake flow behind an airfoil.

Finally, we explore the compression properties of the above constructed divergence-free wavelet bases. Here the term compression refers to the reduction of data sets as mentioned in the Preface on p. VII and should not be confused with the flow property compressibility. We compress several turbulent flow fields in the sense that we retain possibly few terms in their wavelet expansion while keeping most of the energy. Such a reduction without loosing essential information is closely related to the *regularity* of the underlying field in a certain scale of Besov spaces [66]. Due to the norm equivalences (1.26) and (1.28) on p. 9 induced by wavelet expansions we are able to quantify the regularity of given data sets in the relevant scale and make conclusions about their compressibility or *adaptive recovery*, see [28]. This latter aspect is particularly interesting with regard to adaptive direct numerical simulation.

3.2.3 Proper Orthogonal Decomposition (POD)

In general, turbulent shear flows contain discernible, self-sustaining components, called the *coherent structures*, which account for a major part of the turbulent kinetic energy in these flows. Hence, if these structures can be represented by a set of linearly independent functions, this set may be seen as a natural basis for the relevant flow fields. One prominent example of such a basis can be constructed by the *Proper Orthogonal Decomposition (POD)* (or *Karhunen-Loève decomposition*) method, [87, 102], which we now outline.

Basic Idea. In order to describe the main idea behind the POD, we follow [87]. Let us suppose we are a given an ensemble $\{u_k\}_{k=1,\ldots,M}$ of complex valued $L_2(0,1)$-functions. In this context, one wants to find a basis $\{\varphi_j\}_{j=1}^{\infty}$ of $L_2(0,1)$ that is optimal for the data set in the sense that finite-dimensional representations of the form

$$\sum_{j=1}^{N} a_j \, \varphi_j$$

describe 'typical' members of the ensemble better than representations of the same dimension in *any* other basis. The notion 'typical' indicates the use of an averaging operation, which we will denote by $\langle\cdot\rangle$ and which is assumed to commute with the integration over $[0,1]$, i.e.,

$$\int_0^1 \langle \{u_k\}_k \rangle = \left\langle \left\{ \int_0^1 u_k(x)\, dx \right\}_k \right\rangle .$$

Averaging could be done over a number of separate experiments forming $\{u_k\}_{k=1,\dots,M}$ or a time average as we will see below.

With *optimal* it is meant that φ is chosen in order to maximize the averaged projection of a function u onto φ, suitably normalized, i.e.,

$$\sup_{\varphi \in L_2(0,1)} \frac{\langle |(u,\varphi)_{0,(0,1)}|^2 \rangle}{\|\varphi\|^2_{0,(0,1)}}. \tag{3.28}$$

Of course (3.28) would yield only the best approximation to ensemble members by a *single* function, but the other critical points of this functional are also physically significant, since they correspond to a set of functions which, taken together, provide the desired basis.

In order to extremize $\langle |(u,\varphi)_{0,(0,1)}|^2 \rangle$ subject to the constraint $\|\varphi\|^2_{0,(0,1)} = 1$, the corresponding functional for this constrained variational problem is

$$J[\varphi] := \langle |(u,\varphi)_{0,(0,1)}|^2 \rangle - \lambda(\|\varphi\|^2_{0,(0,1)} - 1)$$

and a necessary condition for extrema is that the derivative vanishes for all variations. Straightforward calculations show that

$$\begin{aligned} \frac{d}{d\delta} J[\varphi + \delta\psi]|_{\delta=0} &= 2\,\mathrm{Re}[\langle (u,\psi)_{0,(0,1)}\, (\varphi,u)_{0,(0,1)} - \lambda(\varphi,\psi)_{0,(0,1)}] \\ &= 2\,\mathrm{Re}\left[\int_0^1 \left(\int_0^1 \langle u(x)\bar{u}(y)\rangle \varphi(y)\, dy - \lambda\, \varphi(x) \right) \bar{\psi}(x)\, dx \right] . \end{aligned}$$

Thus, the condition $\frac{d}{d\delta} J[\varphi + \delta\psi]|_{\delta=0} = 0$ reduces to

$$\int_0^1 \langle u(x)\bar{u}(y)\rangle \varphi(y)\, dy = \lambda\, \varphi(x). \tag{3.29}$$

Hence, the optimal basis is given by the eigenfunctions $\{\varphi_j\}$ of the integral equation (3.29) whose kernel is the averaged autocorrelation function $R(x,y) := \langle u(x)\bar{u}(y)\rangle$. They are also sometimes called *empirical eigenfunctions.*

It can be shown that the averaged autocorrelation function has a decomposition

$$R(x,y) = \sum_{j=1}^{\infty} \lambda_j\, \varphi_j(x)\, \bar{\varphi}_j(y),$$

where λ_j are the eigenvalues and φ_j are the L_2-mutually orthogonal eigenfunctions. Finally it can be shown that the latter equation and (3.29) is necessary and sufficient for (3.28). Ordering the eigenvalues $\lambda_j \geq \lambda_{j+1} (\geq 0)$ gives then rise to the Proper Orthogonal Decomposition, i.e.,

$$u_k = \sum_{j=1}^{\infty} a_{k,j}\, \varphi_j. \tag{3.30}$$

POD of Instationary Flows. Let us consider a specific flow scenario in a domain $\Omega \subset \mathbb{R}^n$ for a time interval $[0,T]$. We assume that we are given velocity fields $\boldsymbol{u}_j(x) := \boldsymbol{u}(x,t_j) : \mathbb{R}^n \to \mathbb{R}$, $x \in \Omega$, which can be seen as 'snapshots' of the flow at a time $t_j \in [0,T]$, $j = 1,\dots,N$. It is assumed that this ensemble $\{\boldsymbol{u}_1,\dots,\boldsymbol{u}_N\}$ is chosen in such a way that it 'properly' represents the dynamic of the flow. The averaging is here to be understood as a time average in the sense

$$\langle \boldsymbol{u} \rangle := \frac{1}{N}\sum_{j=1}^{N} \boldsymbol{u}_j$$

which is a vector field (at least) in $\boldsymbol{L}_2(\Omega)$.

Then, the POD method produces an orthogonal set of vector fields $\{\boldsymbol{\phi}_j, j = 1,\dots,\infty\}$ which in the above sense are best correlated with the field. The first N POD modes $\{\boldsymbol{\phi}_i : i = 1,\dots,J \le N\}$ are usually interpreted as capturing the coherent structures of the flow. These (finitely many) modes are often subsequently used as test and trial functions in a spatial Galerkin discretization of the Navier-Stokes equations which results in a system of ordinary differential equations. Thus one obtains a very economical approximate solution that can be used to further study flow characteristics for the given environment for comparisons with other simulations. Our aim here, however, is to use divergence-free wavelets in order to analyze the structure of the POD.

3.2.4 Numerical Implementation and Validation

Implementation. In this section, we describe the numerical implementation of the algorithms needed to carry out a wavelet analysis along the lines indicated above. The underlying code was written in MATLAB. In addition, C++ routines from the MULTISCALE LIBRARY (MSLIB), [6] were used.

The filters for the divergence-free and its complement wavelet transform were computed according to the relations (2.1), (2.2) on p. 84 and (2.48) on p. 99. Based on this, we implemented two- and three-dimensional versions of the divergence-free and the complement wavelet transform. For visualization purposes, only the two-dimensional results will be presented.

In the simplest form described above the divergence-free wavelets are defined on all of $\mathbb{R}^n$ while the computations have to be confined, of course, to bounded domains. To minimize technical complications caused by divergence-free wavelets near domain boundaries we devised two strategies. The first one is based on using the periodized form of the transform on periodic data of a reasonable size. The second one exploits the fact that the data nearly vanish outside a sufficiently large cube. Thus, we simply extend the data with zeros outside the boundaries, so that the finite difference operations given in (2.1) can be applied without any complications.

To study the possible range of data reduction (wavelet compression) it is important to test different filters giving rise to different orders of cancellation properties and ranges for the norm equivalences. Therefore the test data has to be brought into a filter independent format. We used the following approach. We compute first the piecewise linear interpolant for the given discrete sample values. Of course, the resulting function has an *infinite* expansion in terms of the divergence-free wavelets. Therefore we compute next the biorthogonal projection onto a multiresolution space of sufficiently high level. This level is chosen so that the maximum of the relative errors at the grid points stays below 0.2 %.

Validation. We have used two kinds of data for our tests. First, as a basis for comparison, we used data for a periodic flow in a box which has kindly been provided to us by Kai Schneider, see [73]. Then, we used several snapshots of a two-dimensional velocity field obtained from the *Particle Imaging Velocimetry (PIV)* measurements of the turbulent wake behind an airfoil shown in Fig. 3.1 with the parameters displayed in Table 3.1. This data has been kindly made accessible to us by M. Glauser and C.S. Yao from NASA Langley Research Center. Of course, only for low Mach numbers, as in the present case, air flow can be assumed to be incompressible. Moreover, when taking two dimensional slices one cannot expect to obtain strictly divergence-free fields in the turbulent flow regime, hence this data is particularly suitable for the analysis of compressible components in a given flow.

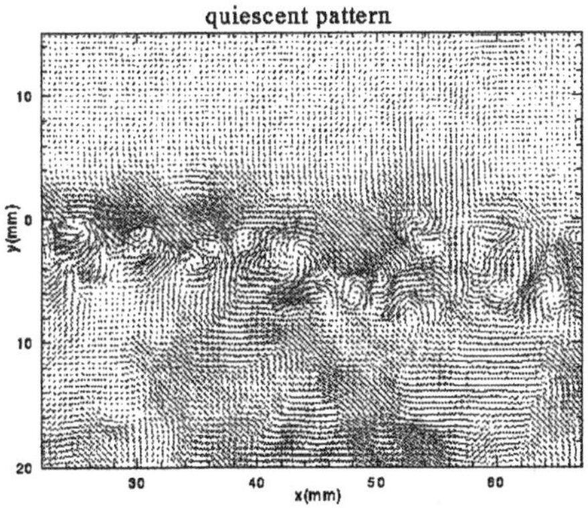

Fig. 3.1. Pattern of the PIV data.

After validating our code with the data reported in [73], we tested the separation of a field obtained from the PIV data into its solenoidal and compressible parts. Such a separation is shown in Fig. 3.4 on p. 134 below. The

Table 3.1. Parameters of our test data for the turbulent wake behind an airfoil.

Experimental setup	
Airfoil	Chord length= 101.6 mm, thickness = 12.7 mm, angle of attack = 4.5 deg.
Wind tunnel	Height = 108 mm, width = 254mm, (closed loop).
Flow	Free stream velocity=18.1m/s, mach number=0.05, Reynolds number= 135 000.
Other Specifications	Airfoil mounted 50.8 mm above the floor of wind tunnel. PIV pictures taken 12 mm downstream of trailing edge.
Grid specifications	
No. of grid points	x: 168; y: 94
First grid point	(21.304 mm, -21.9225 mm)
Last grid point	(100.432 mm, 22.3515 mm)
Grid spacing	0.471 mm

sum of these parts was accurate to double precision with respect to the original field in the L_2-norm independent of the choice of filters which confirms the quantitatively good numerical stability of the multiscale transforms for all filters under consideration. In addition, the divergence of both parts were computed. For the divergence calculation, a second-order central difference scheme on the point values of the chosen field was used in the non-periodic case. These point values were obtained by reconstructing the single scale representation and evaluating the B-splines on a four times refined grid. In the periodic case, we used Fourier techniques. For the PIV data, the divergence of the solenoidal part was at most of order 10^{-5} whereas that of the compressible part was of order 10^1. For the periodic data, our results agree with all the relevant quantities in [73].

3.2.5 Numerical Results I: Data Analysis

In this section, we summarize our numerical results concerning the application of the wavelet decomposition for analyzing several data. First, we present results on a visualization technique that helps to distinguish between regions of compressibility and incompressibility in a flow. Then, we present preliminary results on the divergence-free wavelet analysis and wavelet coefficient visualization of the PIV data. Finally, we perform extensive analysis on the POD modes extracted from the PIV data to gain more insight into the properties of the coherent structures. In particular, we investigate the quantitative influence of artificial compressibility, that stems from the two-dimensional measurement of the available three-dimensional turbulent wake data, on the coherent structures and POD-based incompressible flow modeling. In addition, we study the multiscale structure of the POD modes.

Resolution of Local Effects. In many application especially in image and signal processing, wavelet coefficients are visualized in pseudocolor plots over the various levels, see also Fig. 3.5 on p. 135 below. With these plots, the dominant wavelet coefficients of the multiresolution analysis can be identified. Moreover, a visualization technique in which the coefficients are compiled at the supports of their corresponding functions can also offer insight into supersonic turbulent flows which are, however, not studied in this paper. Given a flow field with localized regions of compressibility, plots of the wavelet coefficients corresponding to Ψ^{dc} can be used to identify these regions at different scales of the flow field. In a similar fashion, the divergence-free wavelet coefficients can be plotted to identify the regions of incompressibility. Moreover, in view of (1.26) on p. 9 and Proposition 5 on p. 100 these plots also indicate regions of low and high energy.

In order to perform the proposed visualization, the support of the basis functions associated with each coefficient has to be computed. As this operation has to be performed at each level for $2^n - 1$ sets of data points, it is computationally expensive. Instead we used an inverse wavelet transform (see Chap. 1) like algorithm, in which the non-zero valued coefficients of all the filters were replaced by ones with the appropriate normalizations. This way, the relative locations of the the different wavelets could be computed on the coefficient level. This procedure has been used to produce pseudocolor plots which display the interplay of different scales in the decomposed field. Additionally, the mapping corresponding to the finest level is used to identify localized regions where a specific wavelet *type* is dominant. To be more precise, this plot corresponds to the sum of all coefficient plots for the different levels. In Figs. 3.2 and 3.3 on p. 132 such plots are shown for two different vector fields. Fig. 3.2 corresponds to an incompressible field with some localized areas of compressibility that are clearly represented by the corresponding divergence-free wavelet coefficients. Finally, in Fig. 3.3, we show the decomposition of one snapshot of the PIV data. The flow structure is also displayed in the wavelet coefficients for both the compressible and incompressible parts.

Analysis of the PIV Data. As a first step, we decompose the flow fields in order to determine their incompressible and compressible parts. The flow fields in Fig. 3.4 on p. 134 show a subsection of one of these snapshots. As our data consists of 2D streamwise slices of the corresponding 3D incompressible turbulent flow field, it is not surprising that the 2D vector fields are not divergence-free. For 240 different realizations of the flow field, the incompressible part accounts for 93% of the total energy on average. In Table 3.2 on p. 126, we have displayed the energies on the different levels of two different snapshots. It can be observed that the loss of energy occurs mostly on the finer scales of the flow, for which the cross-stream turbulent flow fluctuations are important. The divergence-free transform offers a natural means to quantify the associated experimental measurement error over the scales of the flow.

Standard plots of wavelet coefficients over the different levels of the wavelet decomposition offer insight into the energy distribution over the scales of the flow. However, as shown in Fig. 3.5 on p. 135, top, the contributions on higher scales decay so rapidly that they are no longer visible in the plot. To show the pattern of these high scale contributions the coefficients are amplified to attain essentially the same order of magnitude as the coarse scale contributions. In this way an approximate self-similarity is visible, see Fig. 3.5, bottom.

Table 3.2. Energies of the original field and the divergence-free part for different levels.

	Snapshot 15			Snapshot 140		
j	orig.	div-free	perc.	orig.	div-free	perc.
4	0.0070	0.0063	90.4%	0.0070	0.0063	90.4%
3	0.0148	0.0133	89.6%	0.0134	0.0118	87.9%
2	0.0238	0.0222	93.1%	0.0228	0.0212	92.9%
1	0.0383	0.0363	94.6%	0.0343	0.0316	92.4%
0	0.0372	0.0347	93.3%	0.0418	0.0400	95.8%

Analysis of POD Modes. As already mentioned, POD modes may be interpreted as the coherent structures, i.e., those structures which capture the 'main features' of the flow. The divergence-free wavelet analysis offers a tool to gain insight into at least two interesting issues on POD:

Influence of Artificial Compressibility on the POD modes. In many cases, the numerical simulation of turbulence starts with an extensive analysis of a set of the given flow data for carrying out an appropriate turbulence simulation. The kind of flow configuration that we investigate here is usually modeled by the *incompressible* Navier-Stokes equations. Thus, the POD modes extracted from the data are used as test and trial functions in a Galerkin scheme for constructing a low-order numerical approximation of the full set of the Navier-Stokes equations. This in particular imposes the divergence-free requirement on the POD modes. In turn, for the POD functions to be incompressible, the underlying flow data has to also meet the divergence-free condition. However, data obtained from experiments or direct numerical simulations may contain errors due to inexact measurements or numerical effects. These errors may cause the flow field under consideration to be no longer divergence-free. The use of POD modes extracted from such data as a divergence-free basis in an incompressible simulation constitutes a violation on the used model. The Helmholtz decomposition based on divergence-free wavelets offers a tool for quantifying the associated error.

As two-dimensional slices of the PIV data include a naturally built-in artificial compressibility, we chose to use these data for our analysis. To this end, we first created a new data set of divergence-free fields which were obtained by the divergence-free projection. Then we applied POD decomposition on both sets, namely the original and the divergence-free part.

In a first step, we compare the turbulent kinetic energies associated with the decomposed fields using the POD coefficients of the two data sets. From the stability of the POD method, it is expected that the energies of the original and divergence-free field are asymptotically equal. Here, we are interested in quantitative results. We find exactly the same percentage of energy recovery (93%) as mentioned above when we compare the norms of the POD coefficients corresponding to the original and filtered (i.e., divergence-free) data sets.

Next, we are interested in quantitative results concerning the turbulent kinetic energy. The bottom picture in Fig. 3.6 on p. 136 shows the percentage of turbulent kinetic energy captured by each mode for the original field and its divergence-free part. It can be seen that the two curves are quite close. Moreover, we investigate the relative error of the POD approximation of with respect to turbulent kinetic energy. To be more precise, we consider $||\mathbf{v}||_{0,\Omega}$ (i.e., the turbulent kinetic energy of a vector field $\mathbf{v}$) and let $\mathbf{u}$ be the original ensemble, $\mathbf{u}^{\nabla}$ the divergence-free part and $\mathbf{u}_i$, $\mathbf{u}_i^{\nabla}$ the approximations of $\mathbf{u}$ and $\mathbf{u}^{\nabla}$ by the first i POD modes (i.e., $\mathbf{u}_N = \mathbf{u}$, $\mathbf{u}_N^{\nabla} = \mathbf{u}^{\nabla}$). Then the upper plot in Fig. 3.6 shows

$$r_i := \frac{\| \mathbf{u} - \mathbf{u}_i \|_e}{\| \mathbf{u} \|_e}, \qquad r_i^{\nabla} := \frac{\| \mathbf{u}^{\nabla} - \mathbf{u}_i^{\nabla} \|_e}{\| \mathbf{u}^{\nabla} \|_e}$$

over the number of modes, i. The curve for the incompressible part is slightly above the one for the original data. For this kind of experiment, one expects the artificial compressibility to be some kind of high scale noise since the flow is close to two-dimensional. Our results confirm this expectation quantitatively.

Multiscale structure of POD modes. There has been a lot of debate about whether the coherent structures have multiscale structure or not. A wavelet-based multiscale analysis of the POD modes interpreted as the coherent structures can shed some light on this question. For this purpose, we have performed vector wavelet and divergence-free wavelet transforms on several POD modes of the PIV data and its divergence-free part. For each mode starting from the most to the least energetic one we have computed the energy distribution over the different scales of the flow. The results for the two data sets were compiled in Table 3.3 and Table 3.4 on p. 128. In all columns the maximal value is printed in boldface. The results indicate that the energy for the first (i.e., most energetic) few modes is concentrated around the 5th level of the decomposition. For the less energetic modes, the energy is distributed more evenly between the finer scales of the flow. Assuming that the coherent structures are represented by the first 25 POD eigenfunctions which capture about 60% of the turbulent kinetic energy, it can be seen that about 6 scales contribute most to these structures. Hence we see that at least for our data, the coherent structures have multiscale structure.

A final remark can be inferred from the above study: Our preliminary analysis reveals that the deviation from incompressibility is essentially limited to the finer scales of the flow, while the multiscale analysis of the POD modes indicates that the coherent structures are associated with the coarser scales of the flow. Hence the compressibility error induced in the construction of a low dimensional, incompressible flow approximation based on POD is far less significant due to the use of only the first few dominant modes that represent the coherent structures.

Table 3.3. Distribution of L_2-norm energy over scales of the POD eigenfunctions corresponding to the original PIV data .

j/Mode	1	10	25	50	100	200	240
9	0.0628	0.0771	0.0956	0.1186	0.1596	0.2439	**0.8838**
8	0.0885	0.1409	0.1959	0.2527	0.3755	0.5450	0.4662
7	0.1703	0.2322	0.3172	0.4929	**0.6116**	**0.6323**	0.2340
6	0.5583	0.4035	**0.6266**	**0.6135**	0.5316	0.4422	0.1264
5	**0.6961**	**0.6551**	0.5527	0.4307	0.4292	0.2710	0.0637
4	0.1267	0.3596	0.3346	0.3167	0.1627	0.1339	0.0216
3	0.1058	0.2813	0.1661	0.0889	0.0924	0.0839	0.0090
2	0.0790	0.1754	0.0143	0.0361	0.0458	0.0133	0.0022
1	0.0461	0.0504	0.0044	0.0064	0.0038	0.0043	0.0004
0	0.0142	0.0182	0.0006	0.0182	0.0011	0.0011	0.0001

3.2.6 Numerical Results II: Complexity of Turbulent Flows

In this section we use the strong analytical properties of wavelets in order to investigate the complex structure of a turbulent flow. We are interested in two issues. First we study how sparse wavelet expansions of typical flow instances are, i.e., how many degrees of freedom are needed to obtain a certain energy recovery. Second, we investigate how effective an adaptive wavelet method can be for simulating an incompressible turbulent flow.

Table 3.4. Distribution of L_2-norm energy over scales of PIV divergence-free part POD eigenfunctions.

j/Mode	1	10	25	50	100	200	240
7	0.0609	0.0618	0.0779	0.1117	0.1654	0.2630	**0.8556**
6	0.1037	0.1222	0.1630	0.2567	0.3871	0.5518	0.4416
5	0.1513	0.2244	0.3768	0.5006	**0.6154**	**0.6569**	0.2479
4	0.4889	0.4615	**0.5697**	**0.6142**	0.5524	0.3801	0.1087
3	**0.7712**	**0.5571**	0.5058	0.3797	0.3153	0.1725	0.0428
2	0.1868	0.5503	0.3788	0.3335	0.1863	0.1324	0.0146
1	0.0923	0.1968	0.1885	0.1166	0.0807	0.0361	0.0132
0	0.0216	0.0194	0.0882	0.0019	0.0096	0.0063	0.0005

Data Compression. In order to investigate the quantitative aspects of the compression properties of divergence-free wavelets, we decomposed several snapshots of the given flow ensemble. The most frequently used compression technique for wavelets is thresholding. Recall that our method here is based on the norm equivalences (1.26) on p. 9 and Proposition 5 on p. 100 and a close relationship to the nonlinear best N-term approximation (see [28]).

The compression rate can be detected from the relation of the number of coefficients to the retained energy of the field. Fig. 3.7 is a typical example of the compression performance of divergence-free wavelets on the snapshots of the PIV data. 99% of the energy in the L_2-norm can be recovered by dropping 97% of the wavelet coefficients. On the other hand the same recovery in the H^1-norm requires keeping 10% of the coefficients. Let us mention that the compression rates for the periodic data are even higher and are comparable to those reported in [73]. The results displayed in Fig. 3.5 on p. 135 and 3.7 on p. 136 have been computed with the basis corresponding to $d = \tilde{d} = 3$ but we observe similar curves also for other choices of these parameters. This is related to the smoothness of our data which will be investigated in the next subsection.

Potential for Adaptivity. A central objective for further research is to use divergence-free wavelets for a fully adaptive numerical simulation of turbulent flows. The corresponding spatial Galerkin projection of the Navier-Stokes equation could be done with respect to trial spaces spanned by adaptively chosen divergence-free wavelet bases. Quite recently, the connection between adaptive methods and nonlinear approximation has been analyzed rigorously for a wide class of problems, [28, 29]. In particular, it was shown under which circumstances it can be proven that the use of adaptive methods offers an asymptotic gain of efficiency over uniform refinement strategies. This means that the rate of convergence (relating accuracy and degrees of freedom) is higher for the adaptive than for the non-adaptive method. This is closely related to the regularity of the flow fields in certain smoothness scales which will be explained next. In the context of incompressible flows smoothness is usually measured in the scale of Sobolev spaces H^s, i.e., derivatives are measured in L_2. The Besov spaces $B^s_q(L_p)$ can be used to measure smoothness of order $s \in \mathbb{R}$ in L_p where the additional parameter $q \in (0, \infty)$ permits some additional fine-tuning. Besov spaces arise as interpolation spaces between Sobolev spaces, see e.g., [11, 66]. The key here is that they can also be characterized by weighted sequence norms based on wavelet expansions similar to (1.26) on p. 9, [28, 66]. Now fixing s and decreasing p the spaces $B^s_q(L_p)$ become larger and admit an increasingly stronger singular behavior of its elements. The Sobolev embedding theorem tells us how far p can be decreased so that the spaces remain continuously embedded in L_2, say. The critical value $p = \tau$ is given by the relation

$$\frac{1}{\tau} = \frac{s}{n} + \frac{1}{2}, \tag{3.31}$$

where n denotes the spatial dimension. Fig. 3.8 on p. 137 shows a diagram of these spaces. The horizontal axis corresponds to $1/p$, i.e., the measure of integrability and the vertical axis shows the smoothness, s. The vertical line through $(\frac{1}{2}, 0)$ corresponds to the Sobolev spaces where the smoothness is measured in L_2. The diagonal line shows the Besov spaces with the relation of s and τ given by (3.31). Hence, the space $B^s_\tau(L_\tau)$ is (much) larger than H^s when s, τ are related by (3.31). Moreover, this gap increases when s grows and hence τ decreases. The point is that nonlinear approximation schemes and, in particular, the best wavelet N-term approximation, still allow one to recover elements in the much larger space $B^s_\tau(L_\tau)$ with the same accuracy/degree-of-freedom balance as an element in H^s with the aid of schemes based on uniform refinements. The adaptive scheme in [28] was shown to perform in a certain range asymptotically as well as the best N-term approximation. Thus the gain in efficiency offered by an adaptive technique is the larger the more the norm of the approximated object in H^s exceeds its norm in $B^s_\tau(L_\tau)$. In particular, when the approximated function belongs to $B^s_\tau(L_\tau)$ but not to H^s the gain is even reflected by an asymptotically higher convergence rate. Therefore we are interested in the norms of the flows in the spaces $B^s_\tau(L_\tau)$ when s grows. The fact that these norms can indeed be assessed relies on the above norm equivalences. In fact, it is well-known that when s and τ are related by (3.31) one has [66]

$$\left\| \sum_{j,k \in \mathbb{Z}} d_{j,k}\, \eta_{j,k} \right\|^{\tau}_{B^s_\tau(L_\tau(\mathbb{R}^n))} \sim \sum_{j,k \in \mathbb{Z}^n} |d_{j,k}|^{\tau}, \tag{3.32}$$

this relation being valid for a range of s that depends on the underlying wavelet basis, especially its regularity, see also (1.28) on p. 9. In order to use (3.32) to measure the Besov regularity of flow data recall that the data are discrete and therefore have to be mapped first into a function corresponding to an element of a multiresolution space for a very high level. This function automatically belongs (for any set of coefficients) to a given Besov space as long as the underlying wavelets belong to this space and thus (3.32) holds for the corresponding (finite) array of wavelet coefficients. The same applies to the corresponding Sobolev norm given by (1.26) on p. 9. Thus as long as the wavelets have sufficiently high order the two norms can be evaluated for the given flow data in the corresponding regularity range by computing the weighted discrete norms. Therefore we cannot draw any asymptotic conclusions but can only make quantitative comparisons of the two norms.

We used a vector-valued wavelet decomposition corresponding to $\boldsymbol{\Psi}_{e,i} := \Psi_e \, \boldsymbol{\delta}_i$, $\psi_e(x) := \prod_{\nu=1}^{n} {}_{d,\tilde{d}}\vartheta_{e_\nu}(x_\nu)$, with smooth biorthogonal wavelets ($d = 6$, $\tilde{d} = 8$). The results shown in Fig. 3.9 demonstrate that the Sobolev and Besov norms differ more and more for higher values of s. The Besov norms (although being large) even seem to settle independent of s whereas the Sobolev norm increases rapidly. We obtained similar curves for all snapshots of the PIV data and other choices for d and $\tilde{d}$. This indicates a significant potential for wavelet compression that will be addressed in future research.

As far as we know the above conclusions have no theoretical backup yet. To our knowledge, so far regularity results for the relevant Besov scales have been proven for the elliptic case [38, 41], and for 1D conservation laws [67, 68]. In the elliptic case, the Besov regularity was shown to be about twice the Sobolev regularity. In [68] it was shown that even though the solutions to 1D conservation laws may have no positive Sobolev regularity, the Besov regularity in the relevant scale (3.31) is *arbitrarily* high. In view of the results in [28] this implies that nonlinear or adaptive techniques may converge with arbitrarily high order. So far, no analogous results for the incompressible Navier-Stokes equations are known let alone the effect of turbulence.

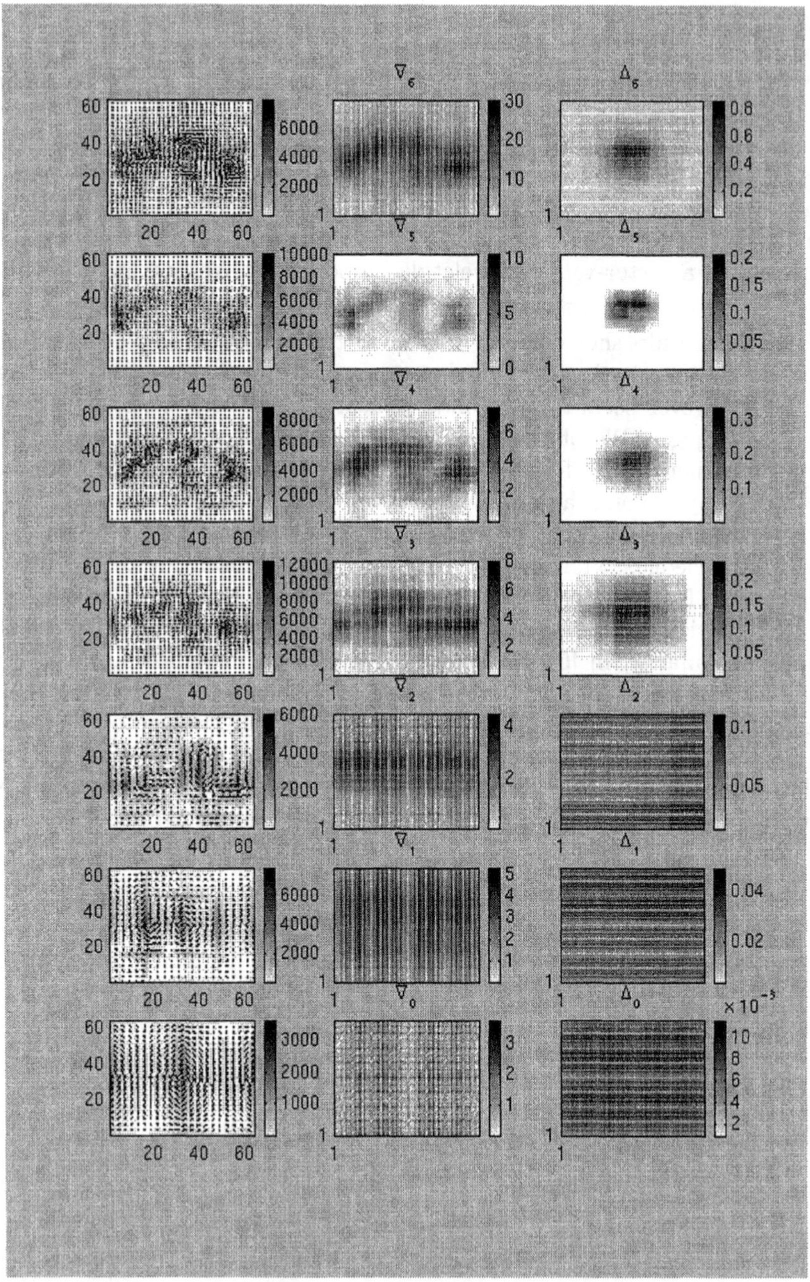

Fig. 3.2. Plot of visualized wavelet coefficients for an incompressible field. Original vector field (left) and wavelet coefficients of the divergence-free part (middle) and compressible part (right), level 6 (top) to level 0 (bottom).

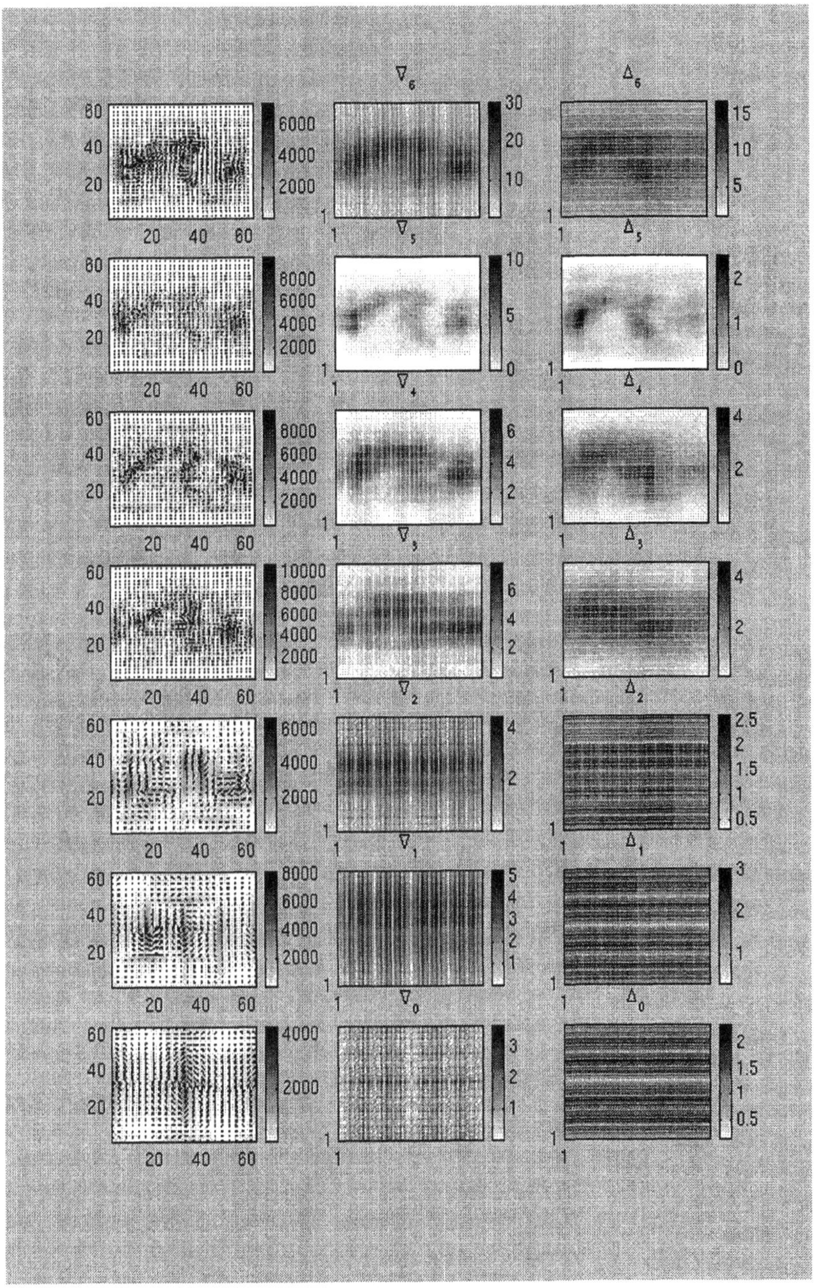

Fig. 3.3. Plot of visualized wavelet coefficients for one snapshot of the PIV data. Original vector field (left) and wavelet coefficients of the divergence-free part (middle) and the compressible part (right), level 6 (top) to level 0 (bottom).

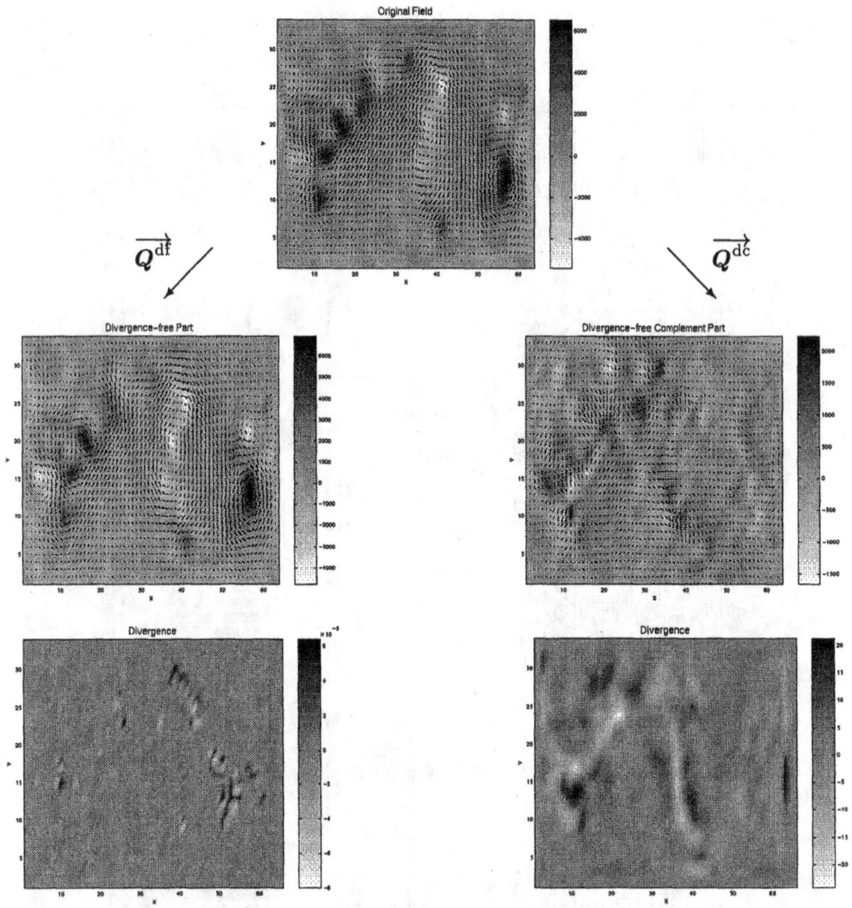

Fig. 3.4. Splitting of a [34X64] section of a snapshot. The first two rows of figures show the velocity superposed with vorticity fields. The last row shows the divergence values.

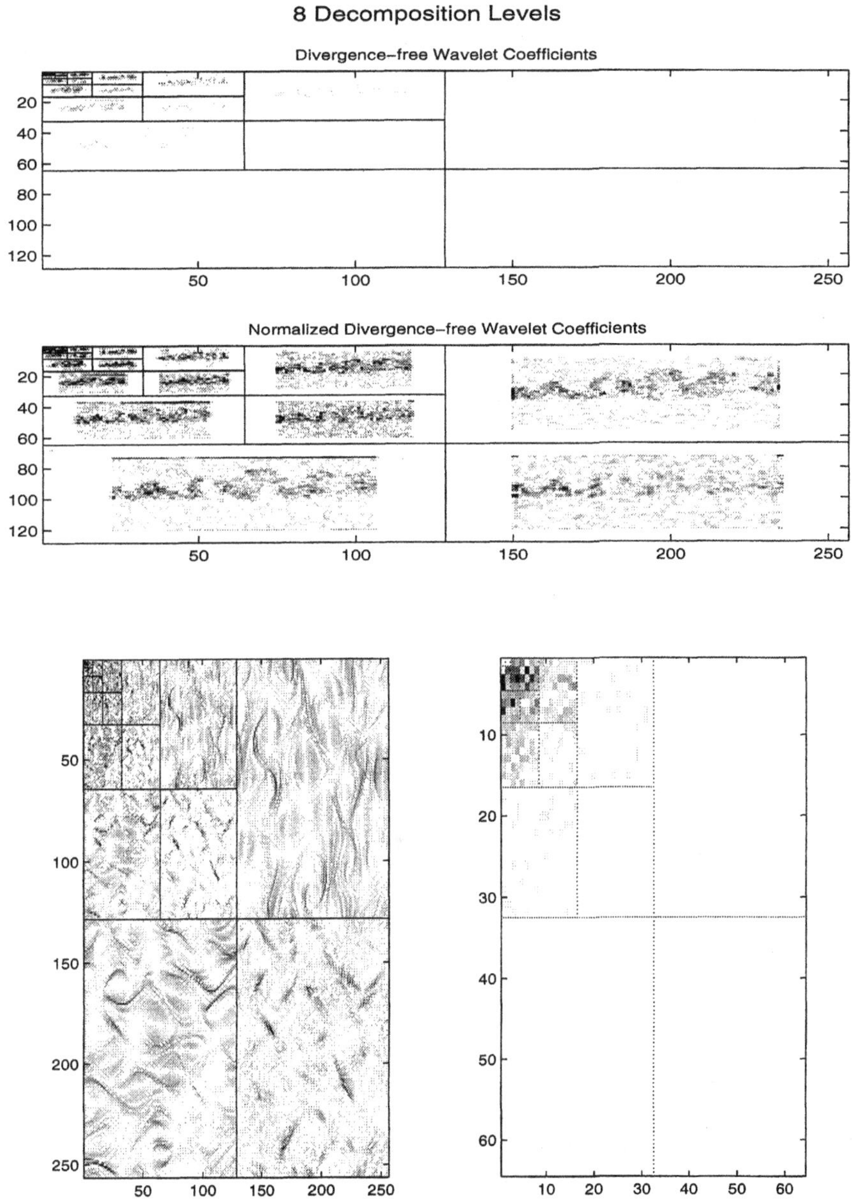

Fig. 3.5. Divergence-free wavelet coefficients of a snapshot plot at coarsening scales of the flow. PIV data (top) and artificial data from [73] (bottom).

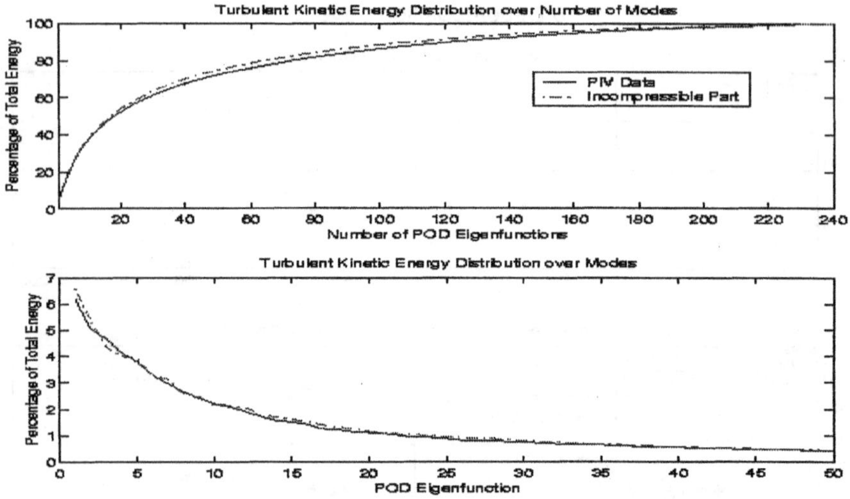

Fig. 3.6. Recovery of turbulent kinetic energy by POD modes for original PIV data and its incompressible part (top). Turbulent kinetic energy per POD mode (bottom).

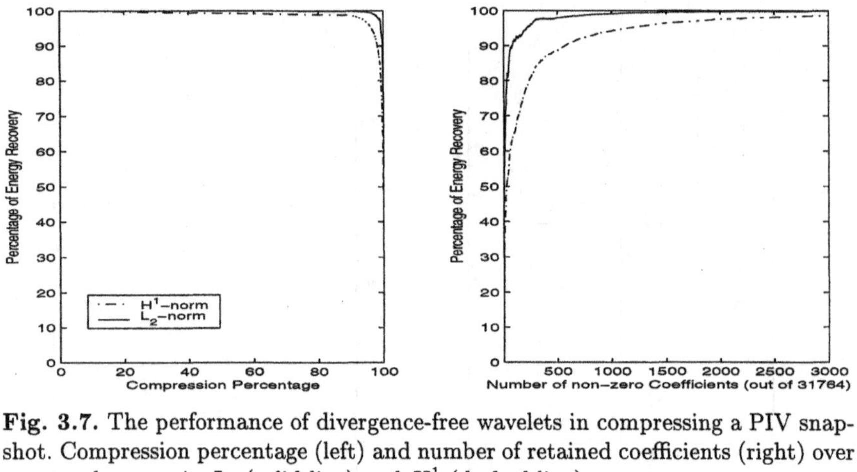

Fig. 3.7. The performance of divergence-free wavelets in compressing a PIV snapshot. Compression percentage (left) and number of retained coefficients (right) over recovered energy in L_2 (solid line) and H^1 (dashed line).

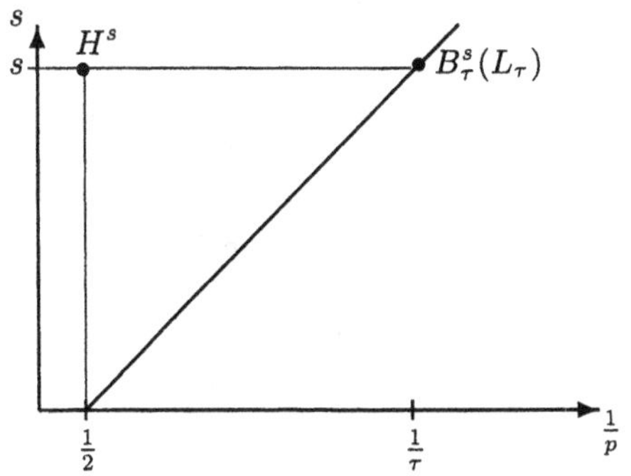

Fig. 3.8. Diagram showing smoothness spaces.

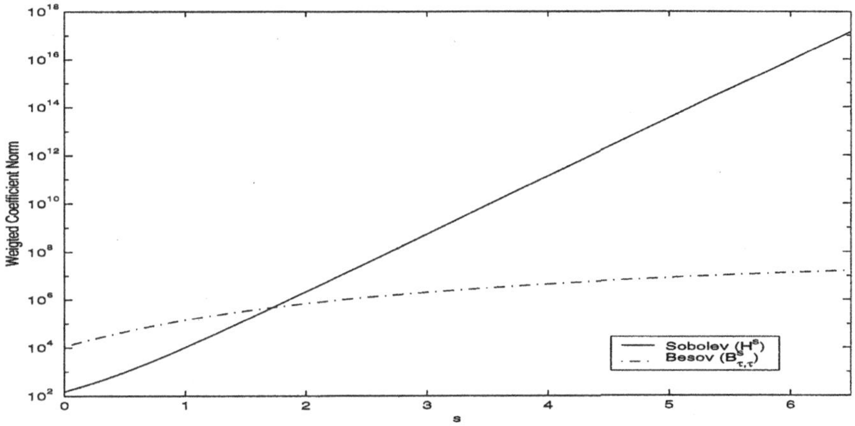

Fig. 3.9. Plots for the Sobolev and Besov regularity of a PIV snapshot in semilogarithmic scale.

3.3 Hardening of an Elastoplastic Rod

Our final example is an elastoplastic problem. The numerical treatment of such problems is still a very challenging field of active research since the governing equations involve several difficulties like extra conditions, local changes of the type of equation, nonlinearities, etc. Besides classical discretization methods such as Finite Differences, Finite Elements and others, first attempts using interpolatory wavelet bases have been made, [35, 83].

A standard technique in instationary numerical elastoplasticity is an elastic predictor-plastic corrector method. At a certain time step, the elastic problem (which is an elliptic one) is solved. The so computed stress is the elastic predictor, which has to be compared with the yielding stresses determined by the material. For certain relations between the stress approximation and the yielding stress, the material changes from elastic to plastic. Hence a plastic correction has to be performed.

The idea is to use a Wavelet-Galerkin method for the elastic predictor and to utilize the good properties, especially the potential for adaptivity of these methods. However, the comparisons with the yielding stress have to be performed *pointwise*. Also possible corrections have to be done pointwise. Hence, we use interpolatory wavelets (see Sect. 1.2.3 on p. 15) in the correction step and perform fast change of bases to switch between the different representations. The results presented here stem partly from [106].

3.3.1 The Physical Problem

We study the dynamic response of a straight elastoplastic rod. In this section, we briefly review the governing equations.

Referring to Fig. 3.3.1, the space variable is denoted by x, the time variable by t and $u(x,t)$ is the axial displacement of the rod. The physical problem is governed by classical relations expressing equilibrium between applied forces and induced internal stresses, compatibility between displacements and strains and the nonlinear constitutive law that relates stresses to strains. As

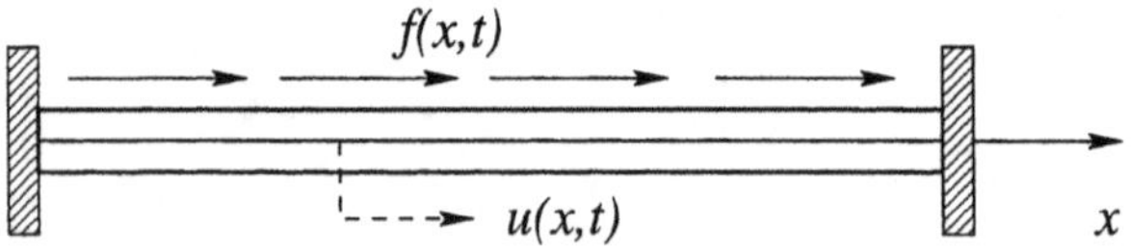

Fig. 3.10. The rod under study

to equilibrium one may write

$$(A\sigma)' + f = \rho \ddot{u}, \tag{3.33}$$

where $A(x)$ and $\rho(x)$ are the cross section and the mass density, $f(x,t)$ is the axial force and σ is the axial stress. Furthermore, space and time differentiation are indicated by a superposed prime and dot, respectively. The hypothesis of small displacement gradients will be made under which the compatibility condition may be written as

$$\varepsilon = u', \tag{3.34}$$

where $\varepsilon(x,t)$ is the axial strain of the rod. The elastoplastic constitutive law classically needs to be introduced in incremental form since the stress does not only depend on the current strain as it happens in the purely elastic case, but also on the entire past history of it. For clarity sake, we first remark that the purely *linear*-elastic problem is governed by a constitutive law that reads

$$\sigma = E_Y \varepsilon, \tag{3.35}$$

in which E_Y is the Young modulus. Therefore, by eliminating stress and strain in (3.33), (3.34) and (3.35), one ends up with a wave equation having the displacement u as unknown, i.e.

$$(E_Y A u')' + f = \rho \ddot{u}. \tag{3.36}$$

We now introduce the basic incremental relationships defining the elastoplastic behavior of the rod. The main hypothesis, widely used and accepted in the literature, [109, 113], is the additive decomposition of the total strain rate $\dot{\varepsilon}$ into its elastic and plastic contributions, i.e.,

$$\dot{\varepsilon} = \dot{\varepsilon}^e + \dot{\varepsilon}^p. \tag{3.37}$$

The stress rate $\dot{\sigma}$ may then be written in terms of any of its above described contributions by introducing the tangent modulus E_t and the plastic one E_p. These are defined by the following relations

$$\dot{\sigma} = E_t \dot{\varepsilon} = E_Y \dot{\varepsilon}^e - E_p \dot{\varepsilon}^p, \tag{3.38}$$

where Fig. 3.11 visualizes the introduced quantities. Notice that in Fig. 3.11, σ_{Y_1} is the so called (tension) *yielding stress*, i.e., the stress above which the material is no longer elastic but undergoes permanent, unreversible deformations. Furthermore, σ_{Y_1} is a variable itself and is to be updated at each time instant according to the so called *hardening rule* that will be discussed next. From a computational point of view, the difficulty is that one does not know in advance whether a stress or strain increment will cause plastic loading or elastic unloading. We assume the rod to be stress-free for $t = 0$ and to behave elastically as long as $(t,x) \in I$ where I is the instantaneous elastic domain defined as

$$I := \{(t,x) \in [0,T] \times (0,1) : -\sigma_{Y_2}(t,x) \leq \sigma(t,x) \leq \sigma_{Y_1}(t,x)\} \tag{3.39}$$

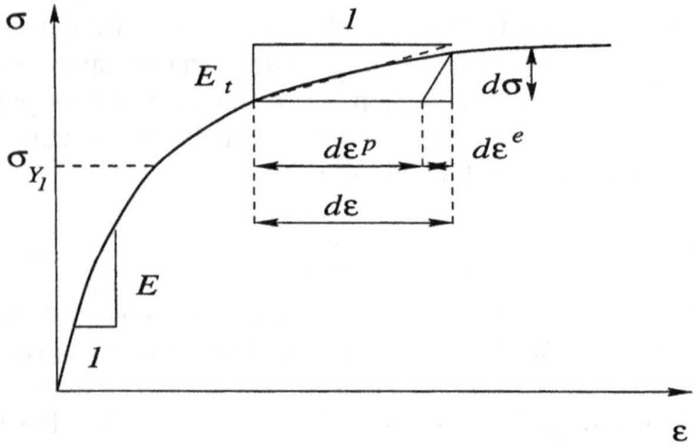

Fig. 3.11. Stress and strain increments

In (3.39), σ_{Y_1} and σ_{Y_2} are the yielding stresses in tension and compression, respectively, which we group into the vector $\sigma_Y = (\sigma_{Y_1}, \sigma_{Y_2})$. For $t = 0$ the yielding stresses are known from experimental tests and they evolve with time following some hardening rule. The key point is that σ_{Y_1} and σ_{Y_2} depend on each other so that it is convenient to separate the total plastic strain into its tension and compression components as

$$\varepsilon^p(t,x) = \kappa_1(t,x) - \kappa_2(t,x) = (1,-1)\begin{bmatrix}\kappa_1\\ \kappa_2\end{bmatrix} =: N\kappa(t,x), \tag{3.40}$$

where $N = [1 \;\; -1]$ and κ_1 and κ_2 are non-negative and non-decreasing. One may then write

$$\dot{\kappa}_i(t,x) \geq 0, \quad \text{for all} \quad (t,x) \in [0,T] \times (0,1). \tag{3.41}$$

The elastic domain in (3.39) is redefined next by introducing the so-called *plasticity functions* $\varphi = (\varphi_1, \varphi_2)^T$ which satisfy the relation

$$\begin{bmatrix}\varphi_1\\ \varphi_2\end{bmatrix} := \begin{bmatrix}1\\ -1\end{bmatrix}\sigma - \begin{bmatrix}\sigma_{Y1}\\ -\sigma_{Y2}\end{bmatrix} \leq \begin{bmatrix}0\\ 0\end{bmatrix},$$

or, in vector form

$$\varphi(t,x) = N^T\sigma(t,x) - \sigma_Y(t,x) \leq 0, \tag{3.42}$$

where the inequality is meant componentwise. Plastic strains may accumulate if and only if both the current stress state and its update belong to the boundary of the elastic domain, i.e., if and only if $\varphi = 0$ and $\dot{\varphi} = 0$. If this is not the case an elastic phase takes place. Therefore, the following complementarity rules govern the entire process

$$\varphi_i \leq 0, \quad \dot{\kappa}_i \geq 0, \quad \varphi_i\dot{\kappa}_i = \dot{\varphi}_i\dot{\kappa}_i = 0. \tag{3.43}$$

By differentiating (3.42), the time rate of the elastic domain may be computed as

$$\dot{\varphi}_i = \pm(\dot{\sigma} - \dot{\sigma}_{Yi}) = \pm\left(\dot{\sigma} - \sum_{j=1}^{2} \frac{\partial \sigma_{Yi}}{\partial \kappa_j}\dot{\kappa}_j\right) = \pm\dot{\sigma} - \sum_{j=1}^{2} H_{ij}\dot{\kappa}_j, \tag{3.44}$$

or

$$\dot{\varphi} = N^T\dot{\sigma} - H\dot{\kappa}, \tag{3.45}$$

where H is the so-called *hardening matrix*. In the case of linear hardening, i.e., when the plasticity functions may be written as

$$\varphi(t,x) = N^T\sigma(t,x) - H\kappa(t,x) - \sigma_Y(0,x), \tag{3.46}$$

one gets the following generalized hardening rule

$$\begin{bmatrix} \sigma_{Y_1} \\ \sigma_{Y_2} \end{bmatrix} = \begin{bmatrix} \sigma_{Y_0} \\ \sigma_{Y_0} \end{bmatrix} + \begin{bmatrix} H_{11} & H_{12} \\ H_{21} & H_{22} \end{bmatrix} \begin{bmatrix} \kappa_1 \\ \kappa_2 \end{bmatrix},$$

or, in vector form,

$$\sigma_Y(t,x) = \sigma_Y(0,x) + H\kappa(t,x). \tag{3.47}$$

The frequently used case of isotropic and kinematic hardening are recovered as particular cases by choosing H respectively equal to

$$H_{\text{iso}} = h\begin{bmatrix} 1 & 1 \\ 1 & 1 \end{bmatrix}, \quad H_{\text{kin}} = h\begin{bmatrix} 1 & -1 \\ -1 & 1 \end{bmatrix}. \tag{3.48}$$

The elastoplastic incremental constitutive law may finally be written as

$$\begin{aligned} \dot{\sigma} &= E_Y(\dot{\varepsilon} - \dot{\varepsilon}^p), \qquad \dot{\varepsilon}^p = N\dot{\kappa}, \\ \varphi &= N^T\sigma - \sigma_Y(\kappa), \quad \dot{\varphi} = N^T\dot{\sigma} - H\dot{\kappa}, \\ \varphi &\le 0, \quad \dot{\kappa} \ge 0, \varphi^T\dot{\kappa} = 0 \quad \dot{\varphi}^T\dot{\kappa} = 0. \end{aligned}$$

Then we obtain our problem under consideration as

Problem 1. For given ρ, E_Y, E_t, E_p, A and f we seek for u, ε and σ such that:

$$\rho(x)\,\ddot{u}(t,x) - \big(E_Y(x)\,A(x)\,u'(t,x)\big)' = f(t,x), \qquad (t,x) \in I, \tag{E}$$

and

$$\begin{aligned} \rho(x)\,\ddot{u}(t,x) - \big(A(x)\,\sigma(t,x)\big)' &= f(t,x), \; (t,x) \notin I, \\ u'(t,x) - \varepsilon(t,x) &= 0, \qquad (t,x) \notin I, \\ \dot{\sigma}(t,x) - E_t(\sigma(t,x))\,\dot{\varepsilon}(t,x) &= 0, \qquad (t,x) \notin I, \end{aligned} \tag{P}$$

Finally, we pose the following initial and boundary conditions

$$u(0,x) = u_0(x), \quad u'(0,x) = u_1(x), \qquad x \in [0,1], \tag{B_1}$$

for some given functions u_0 and u_1, as well as

$$u(t,0) = u(t,1) = 0, \qquad t \in [0,T]. \tag{B_2}$$

3.3.2 Numerical Treatment

In this section, we are going to describe the numerical treatment of Probl. 1. We will detail the elastic predictor-plastic corrector method and we introduce the discretization both in space and time.

Elastic Prediction and Plastic Correction. Let us start by the elastic predictor-plastic corrector strategy. There exist several cases to be handled numerically, all of which are particular cases of the complementarity rule (3.43). For clarity sake we hereafter focus on one of them, i.e. the case of plastic loading or elastic unloading in tension. Let then $\varphi(t,x) = 0$ for some $t \in [0,T]$ and $x \in (0,1)$, meaning that the stress $\sigma(t,x)$ is on the boundary of the instantaneous elastic domain. Given is also $u(t,x)$ for some time t. Then, for a given $\Delta t > 0$, we compute the *elastic predictor* $u^*(t + \Delta t, x)$ by solving the problem (E) (since we always consider the initial and boundary conditions, we will omit referring to them all the time). Then, by using the second equation in (P) (see also (3.34)), we compute ε^* and then, by using (3.35), we obtain the elastic trial stress σ^*. If $\sigma^* < \sigma_{Y_1}$ then an elastic unloading has taken place and therefore $\sigma(t + \Delta t, x) = \sigma^*$ and no correction is required. If conversely $\sigma^* > \sigma_{Y_1}$, a strain-driven modified *Newton-Raphson* correction scheme is used and illustrated in Fig. 3.12. The stress

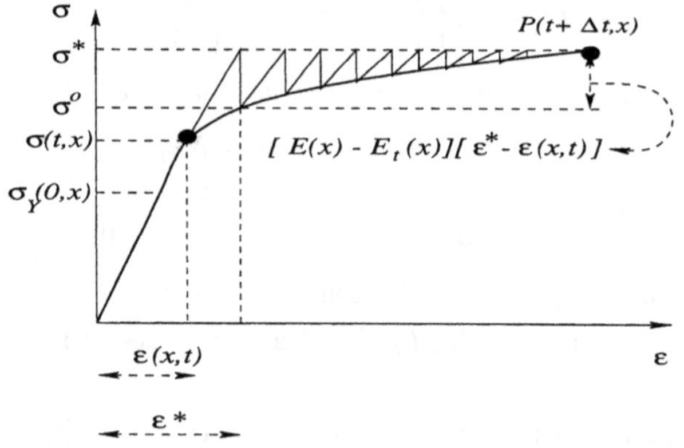

Fig. 3.12. The elastic-predictor plastic-corrector procedure

$\sigma^* = \sigma(t,x) + E_Y(x)[\varepsilon^*(t,x) - \varepsilon(t,x)]$ is no longer updated and will therefore coincide with the value $\sigma(t + \Delta t, x)$. A lack of consistency may however be noticed between the so computed stress σ^* and the one which is compatible with the actual stress-strain curve, i.e., $\sigma^o(t,x) = \sigma(t,x) + E_t(x)\varepsilon^*(t,x)$. The quantity $A(x)[\sigma^* - \sigma^o](t,x)$ becomes a virtual, un-equilibrated force that is brought to the right-hand side of eq. (E) so as to allow the computation of a further update for the displacement and for the strain ε. The procedure ends when the actual stress-strain curve joins the plateau σ^* where the solution in terms of stress, strain and displacement is attained (point $P(t+\Delta t, x)$ in Fig. 3.12). Notice that σ^* is not only the stress $\sigma(t+\Delta t, x)$ but also the new value of the yielding stress σ_{Y_1} to be used for the subsequent stress computation.

When linear isotropic hardening is adopted, see (3.48), the whole stress correction procedure is governed by eight alternative cases. We will hereafter describe the tension cases in detail. The remaining (compression) cases can easily be obtained by symmetry arguments and by replacing σ_{Y_1} by σ_{Y_2}. We will be using discrete-time notations and denote by σ_n the computed stress value at time t_n, by σ^*_{n+1} the trial stress value at time t_{n+1} and by σ_{Y_1}, σ_{Y_2} the current values of the yielding stress. The value of the corrected stress is denoted by σ°.

1. $0 \leq \sigma_n \leq \sigma^*_{n+1} \leq \sigma_{Y_1}$:
   If σ_n and the trial stress σ^*_{n+1} are both below the yielding stress σ_{Y_1}, we are still in the elastic range, i.e., no correction has to be performed. This implies $\sigma^\circ := \sigma^*_{n+1}$.
2. $0 < \sigma_n \leq \sigma^*_{n+1}$, $\sigma_n < \sigma_{Y_1} < \sigma^*_{n+1}$:
   In the case that σ_n is below σ_{Y_1}, but the elastic trial σ^*_{n+1} is above, then there is a transition from the elastic to the plastic regime and the following correction has to be made. Using the third equation in (P), one projects the non-equilibrated part of the stress, namely $\sigma^*_{n+1} - \sigma_{Y_1}$ to the stress-strain curve as can be seen in Fig. 3.13 To be precise, setting $r = (\sigma^*_{n+1} - \sigma_{Y_1})(\sigma^*_{n+1} - \sigma_n)^{-1}$, we obtain $\sigma^\circ = \sigma_{Y_1} + rE_t\varepsilon^p$. Note that the stress difference $\sigma_{Y_1} - \sigma_n$ (and the corresponding strain) still belongs to the elastic range and needs not to be corrected.
3. $\sigma^*_{n+1} \geq \sigma_n \geq \sigma_{Y_1}$, $\sigma^*_{n+1} > \sigma_{Y_1}$:
   In this case, σ_n is in the plastic range and σ^*_{n+1} remains plastic. As in the latter case, the non-equilibrated part of the stress needs to be corrected as indicated by Fig. 3.13, i.e., $\sigma^\circ = \sigma_n + E_t(\varepsilon^p_{n+1} - \varepsilon^p_n)$.
4. $-\sigma_{Y_2} \leq \sigma^*_{n+1} \leq \sigma_n$, $\sigma_n > 0$:
   In this case, unloading is performed and we reenter (or stay in) the elastic regime. Hence, we have to check that σ^*_{n+1} is larger than the compression yielding stress (recall, that $\sigma_{Y_2} > 0$) and no correction is made, $\sigma^\circ = \sigma^*_{n+1}$.

Space Discretization: Wavelet Spaces. For the discretization in space, we use biorthogonal B-spline wavelet systems. We use the wavelet systems $\Psi_{[j]}$ (see (1.33) on p. 10) for generating test and trial spaces for a Wavelet-Galerkin method as in Sect. 3.1.1. Then, we will analyze the wavelet coefficients corresponding to these bases in order to detect regions of plasticity.

Discretization in Time: Newmark Scheme. In this subsection, we briefly review the definition and the main properties of the *Newmark scheme* which we use for the time discretization. Let us mention first, that this scheme

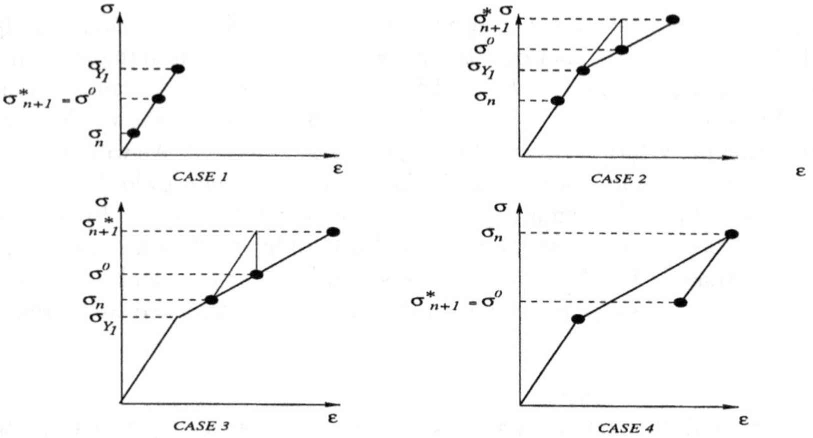

Fig. 3.13. Hardening cases.

is widely used for resolving evolutionary problems in elastic wave propagation and structural mechanics.

We consider a system of differential equations of the type

$$M\,\ddot{X}(t) + K\,X(t) = F(t), \tag{3.49}$$

where the matrices M and K are symmetric, K is positive (semi-)definite and M is positive definite. Then, for a given time step $\Delta t = t_{n+1} - t_n > 0$, the quantities $X(t_n)$, $\dot{X}(t_n)$ and $\ddot{X}(t_n)$ are approximated by X^n, $\dot{X}^n$ and $\ddot{X}^n$, respectively, due to the following relations

$$M\,\ddot{X}^{n+1} + K\,X^{n+1} = F(t_{n+1}), \tag{3.50}$$

$$\dot{X}^{n+1} = \dot{X}^n + \Delta t[(1-\delta)\,\ddot{X}^n + \delta\ddot{X}^{n+1}], \tag{3.51}$$

$$X^{n+1} = X^n + \Delta t\dot{X}^n + \Delta t^2\left[(\tfrac{1}{2}-\theta)\,\ddot{X}^n + \theta\,\ddot{X}^{n+1}\right], \tag{3.52}$$

resulting in the three-level scheme

$$\begin{aligned} &\frac{1}{\Delta t^2}M(X^{n+1} - 2X^n + X^{n-1}) + \\ &+K\left[\theta X^{n+1} + (\tfrac{1}{2}+\delta-2\theta)X^n + (\tfrac{1}{2}+\theta-\delta)X^{n-1}\right] = F^*, \end{aligned} \tag{3.53}$$

with

$$F^* := \theta F^{n+1} + (\tfrac{1}{2}+\delta-2\theta)F^n + (\tfrac{1}{2}+\theta-\delta)F^{n-1}, \tag{3.54}$$

where the real parameters θ and δ have to be chosen in order to guarantee consistency and stability, [64]. Frequently used parameters are $\delta = \frac{1}{2}$ and $\theta \geq \frac{1}{4}$ which results in a second order scheme which is stable for all Δt, [64].

Finite Dimensional Problems. Now, we collect all the above described pieces and end up with a discretization of Probl. 1.

At any time step t_n we will approximate the solution by means of a wavelet expansion of the type

$$u(t_n, x) \sim \sum_{\lambda \in \mathcal{J}_{[j]}} u_\lambda(t) \Psi_\lambda(x) := u_n(x),$$

where $u_\lambda(t)$ are suitable coefficients. Then, the Galerkin approximation of eq. (E) reads

$$\sum_{\lambda \in \mathcal{J}_{[j]}} \left\{ \ddot{u}_\lambda(t)(\rho\psi_\lambda, \psi_\mu)_{0,(0,1)} + u_\lambda(t)(E_Y A \psi'_\lambda, \psi'_\mu)_{0,(0,1)} \right\} = (f, \psi_\mu)_{0,(0,1)},$$

for $\mu \in \mathcal{J}_{[j]}$. Hence, we obtain the system (3.49) with the following parameters

$$\begin{aligned} M &= M_n^\rho &&:= \left((\rho\psi_\lambda, \psi_\mu)_{0,(0,1)}\right)_{\lambda,\mu \in \mathcal{J}_{[j]}}, \\ K &= K_n^{E_Y A} &&:= \left((E_Y A \psi'_\lambda, \psi'_\mu)_{0,(0,1)}\right)_{\lambda,\mu \in \mathcal{J}_{[j]}}, \\ X(t) &= U_n(t) &&:= (u_\lambda(t))_{\lambda \in \mathcal{J}_{[j]}}, \\ F(t) &= F_n(t) &&:= \left((f(t,\cdot), \psi_\mu)_{0,(0,1)}\right)_{\mu \mathcal{J}_{[j]}}. \end{aligned}$$

This means that in each step of the Newmark scheme, we have to solve the linear system

$$\tilde{A}^{n+1} X^{n+1} = \tilde{F}^{n+1},$$

where

$$\begin{aligned} \tilde{A}^{n+1} &:= \tfrac{1}{\Delta t^2} M_n^\rho + \theta K_n^{E_Y A}, \\ X^{n+1} &= U_{n+1}(t_{n+1}), \\ \tilde{F}^{n+1} &= \left(\tfrac{2}{\Delta t^2} M_n^\rho - (\tfrac{1}{2} + \delta - 2\theta) K_n^{E_Y A}\right) X^n \\ &\quad - \left(\tfrac{1}{\Delta t^2} M_n^\rho - (\tfrac{1}{2} + \theta - \delta) K_n^{E_Y A}\right) X^{n-1} - F^*, \end{aligned}$$

where F^* is determined by (3.54).

3.3.3 Stress Correction and Wavelet Bases

In a certain sense, the philosophy of a Galerkin method for the discretization in space contradicts the elastic-predictor plastic-corrector strategy. The latter involves *point values* of the basis functions whereas the Galerkin discretization only uses *integrals* over (products of derivatives of) these functions. The reasons for using a Galerkin approach have been detailed above. However, the pointwise correction is physically adequate. Hence, we performed a mixture of the two methods as described below.

For the discretization in space we use a Galerkin method with biorthogonal B-spline wavelets. When it comes to the stress correction, we perform a change of basis to an interpolatory wavelet basis. The correction can easily be performed with the functions and also the change of basis is not very costly. The resulting unequilibrated force serves as a right-hand side in an elliptic problem. Hence, it is rather easy to perform an L_2-projection of this function in order to determine the entries of the right-hand side. This approach allows us to combine two advantages, namely the use of mathematically founded Wavelet-Galerkin methods with biorthogonal wavelets and an efficient stress correction using interpolatory wavelets.

3.3.4 Numerical Results I: Variable Order Discretizations

We have performed several numerical experiments for the above described problem. Let us mention that the code has been written in C++ using the *MultiLevel Library*, [6]. The code has been validated by the code from [35] and also by comparisons with the standard Finite Element code *FEAP* described in [120]. We will describe here only two main experiments. In the first one, we investigate different discretizations, in particular, we consider wavelet bases of different order and study the interplay between the interpolatory basis and the bases used in the Galerkin method. This will guide the choice of a plasticity indicator which will be described and investigated in the second experiment.

The input data for our test example have been chosen as:

$$\begin{array}{lll} T = 0.4, & \Delta t = 0.01, & \sigma_{Y_1} = \sigma_{Y_2} = 1600, \\ A(x) \equiv 100, & \rho(x) \equiv 7.85, & E_Y = 2100000, \\ E_t = 200000, & f(t,x) := \chi_{[0,0.7]}(x)\, a\, \sin(\omega t), & a = 8 \cdot 10^5, \omega = \frac{20}{3}\pi, \end{array}$$

and within the space domain $x \in (0,1)$. The solution is shown in Fig. 3.14, where on the left we have displayed the displacement and on the right the characteristic stress/strain curve at the two points $x = 0$ and $x = 0.98$. This example is particularly interesting since all hardening cases described in Fig. 3.13 in fact appear. Moreover, in Fig. 3.15, we have plotted the resulting plastic indicators. We obtain two plastic waves starting from the boundaries.

Our first experiment concerns discretizations of different orders, i.e., biorthogonal B-spline wavelets for different choices of the parameter d as described at the end of Sect. 3.3.2. Note that the order of the basis functions used in the Wavelet-Galerkin method and of the basis for the stress correction need not to coincide. Of course, the solution quantitatively is the same as the one indicated by the Figs. 3.14 and 3.15. Here, we want to study some qualitative aspects. In Fig. 3.16, we show data for the solutions concerning $d = 2, 3, 4$ all for level $J = 10$ using piecewise linear interpolatory wavelets up to level $J_{\mathrm{Res}} = 10$ for the stress correction. The first picture in Fig. 3.16 shows the error for $d = 2$. Note that the error is in the range of 10^{-7}. The two remaining pictures in Fig. 3.16 show the difference between the solutions for $d = 2$, $d = 3$ (in the middle) and $d = 2$, $d = 4$ (right). We see that the distance is almost negligible except in two very localized regions near $x = 0$ and

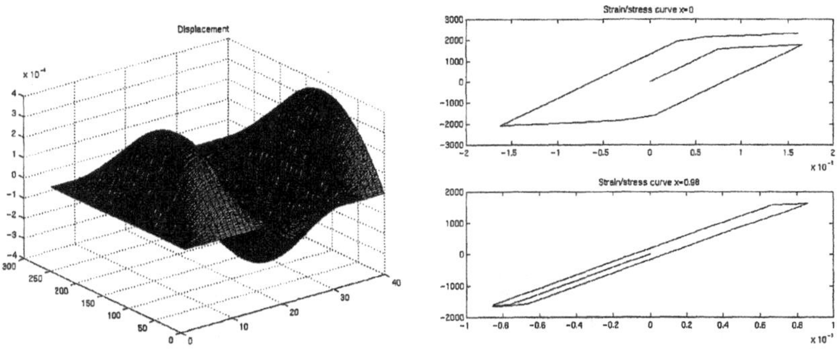

Fig. 3.14. Displacement (left) and characteristic strain/stress curve (right).

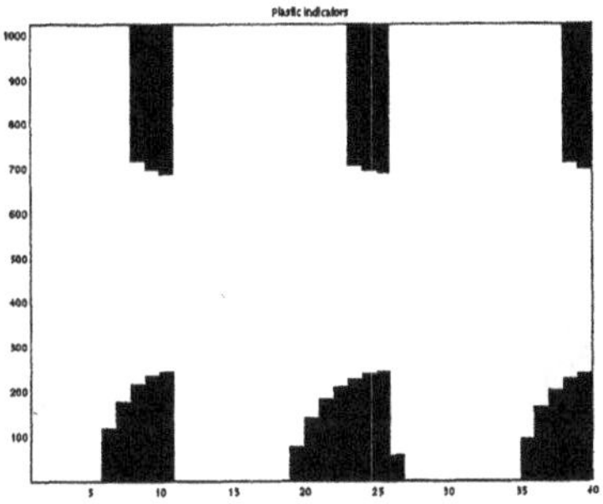

Fig. 3.15. Plastic indicators. The x-axis corresponds to the time steps ($n = 1, \dots, 40$) and the y-axis to the points $i/1024$, $i = 1, \dots, 1024$ in space.

$x = 0.7$. It can be seen from Fig. 3.15 that these are exactly those positions of the plastic wave. It seems that using a *global* high order discretization does

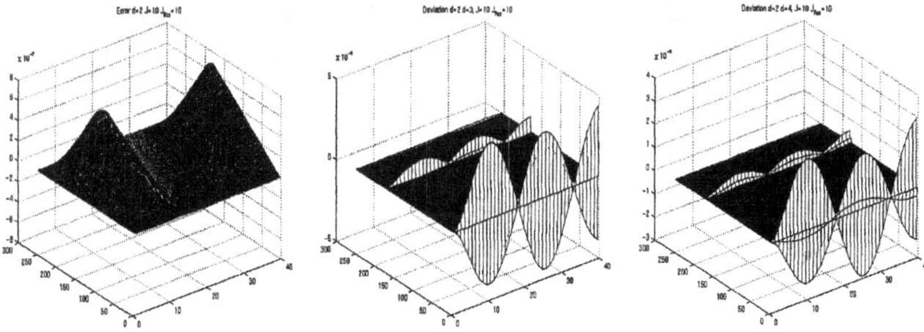

Fig. 3.16. Error for $d = 2$ (left) and difference of the solutions for $d = 2, 3$ (middle) and $d = 2, 4$ (right).

not pay off for this example when simply looking at the differences. However, the detection of regions where plasticity is spreading seems to be improved by high order wavelets. In order to understand this behavior, we consider the parts of the solution corresponding to different wavelet levels. In Fig. 3.17, these parts are displayed for $d = 2$. We see that the details (i.e., the wavelet parts) in fact reflect the beginning of a plastic wave. We now come

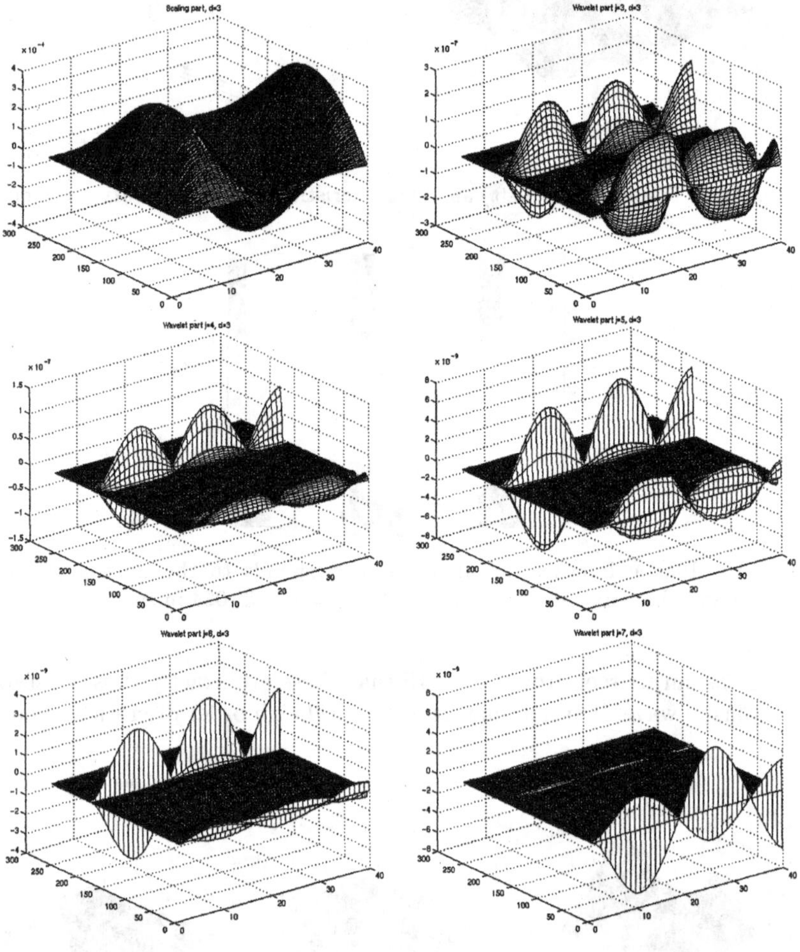

Fig. 3.17. Scaling and wavelet parts of the computed solution for the different levels, $d = 2$.

to the relationship between the wavelet coefficients and plastic phenomena that should justify the use of more advanced wavelet methods. Therefore, in Figs. 3.18 and 3.19, we show the scaling and wavelet coefficients for $d = 2$ and $d = 3$, respectively. Obviously, the behavior of the coefficients for $d = 3$ reflects the occurance of plasticity much better. We see large peaks at the borders of the waves. These pictures show in particular that the loss of high order convergence is (at least partially) due to the interplay of the different wavelet bases in the stress correction. Since the support of the trial functions for $d = 3$ and $d = 4$ is larger than the one of the piecewise linear interpolatory wavelets, the L_2-projection of the unequilibrated force may cause an error.

Note that we keep using *piecewise linear* interpolatory wavelets for the stress correction. This could limit the maximal reachable order, even though the stress correction only effects the right-hand side. Thus, one would have to increase either the level J_{Res} or the order of the interpolatory wavelets. Finally, the nature of the problem itself might also limit the order of approximation. In fact, the regularity theory of such problems is still an open field, [84]. This is the reason why we have to perform numerical tests in order to derive a meaningful plasticity indicator.

In the light of the above result, we confine the next discussions to piecewise linear trial and test functions for the Wavelet-Galerkin method. First of all, we display the error in the Sobolev norms $\|u(t,\cdot) - u_j(t,\cdot)\|_{s,(0,1)}$, $s = 0, 1$ at some time $t \in [0, T]$ for different values of j. Here, u_j denotes the computed Galerkin solution w.r.t. S_j and u is the 'exact' solution. The latter one is computed by using trial and test spaces S_j with a large value for j.

In Fig. 3.20, we have displayed the evolution of these errors in time. We see that the comparably big values of the errors correspond to the occurance of the plastic waves.

Now, we consider two choices for t and monitor the errors in dependence of the maximum level. Our first choice is the final time $t = 0.39$. The two pictures in Fig. 3.21 show the errors in semilogarithmic scale (left) and the rate of convergence (right). We obtain almost the same rate of convergence as for the Wavelet-Galerkin Method applied to an elliptic problem. The same information as in Fig. 3.21 for $t = 0.39$ is displayed in Fig. 3.22 for $t = 0.16$. Note that at this time, no plastic wave occurs in the displacement, see Fig. 3.15, and the general behaviour is similar as in the above case.

In the next experiments, we fix the highest level for the Galerkin discretization to $J = 9$ and consider the L_2-error for different choices of J_{Res}, i.e., the maximum level for the stress correction. In Fig. 3.23, the corresponding error is displayed for $d = 2$ and the choices $J_{\mathrm{Res}} = 3, 4, 5, 6$. We obtain a smoother and smoother error for increasing values of J_{Res}. In the Figs. 3.24 and 3.25, we display the L_2-error and the rate of convergence for $t = 0.16$ and $t = 0.39$. We cannot detect an obvious rate of convergence from these pictures. However, the interplay between the wavelet bases for the Galerkin method and the stress correction seems to be of great importance.

What we can see from this experiment is the reduced rate of convergence in those regions which form the interface between elastic and plastic regions. From Probl. 1, one can in fact expect that the regularity of the displacement function decreases at the interfaces between elastic to plastic regions.

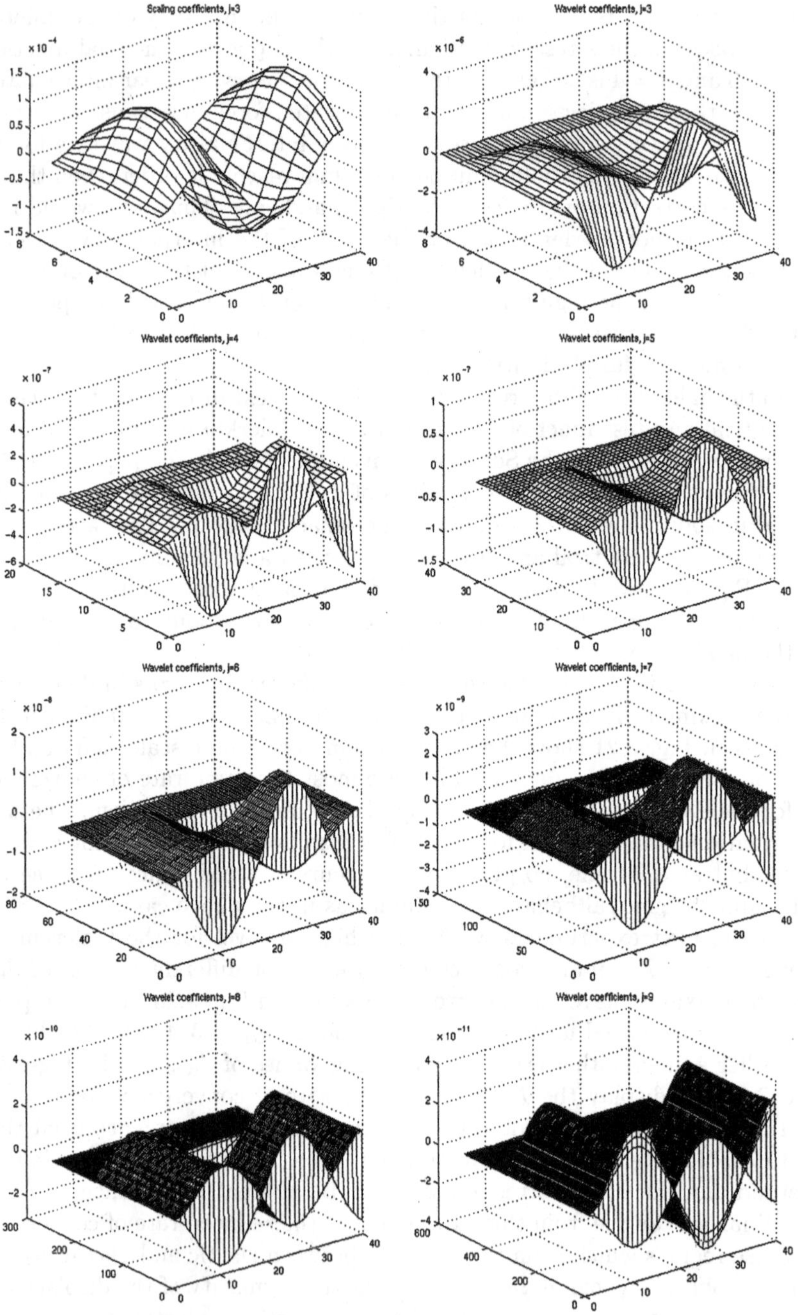

Fig. 3.18. Scaling and wavelet coefficients of the computed solution for the different levels, $d = 2$.

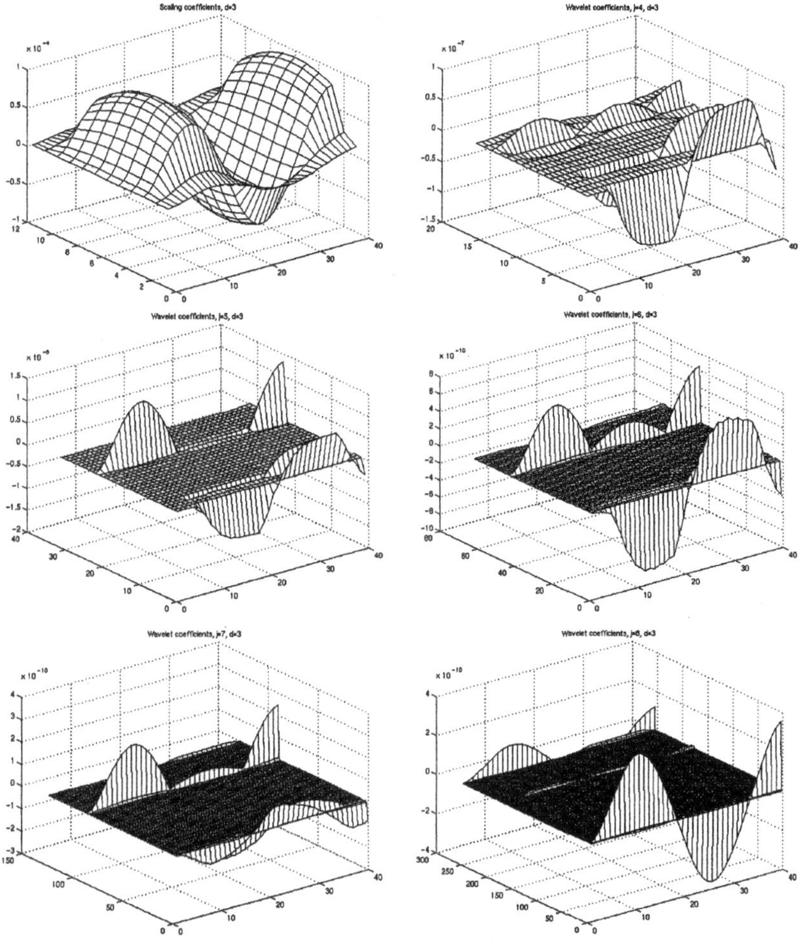

Fig. 3.19. Scaling and wavelet coefficients of the computed solution for the different levels, $d = 3$.

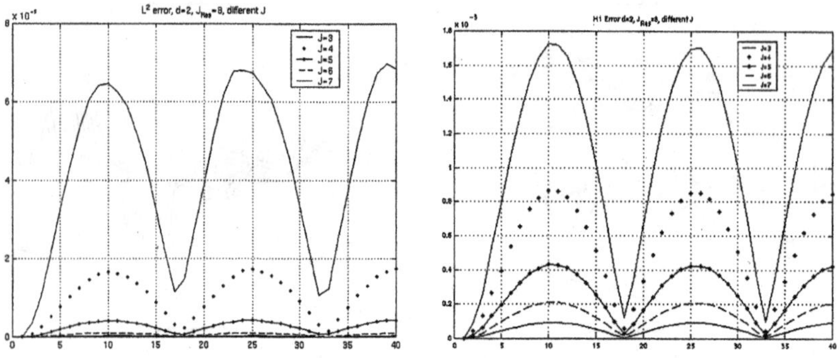

Fig. 3.20. L_2- and H^1-error over t for different maximal levels J.

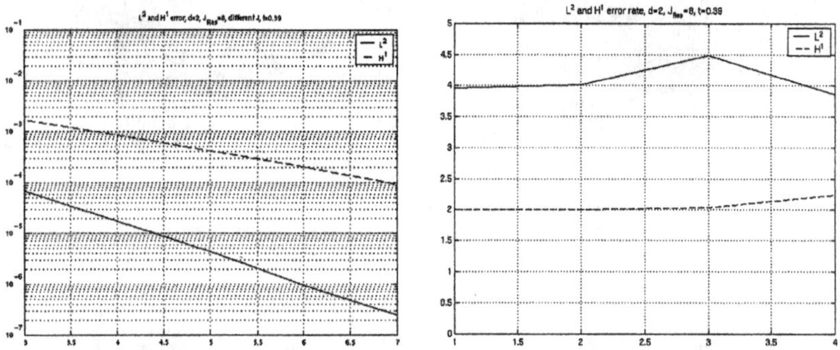

Fig. 3.21. L_2- and H^1-error over the maximal level for $t = 0.39$.

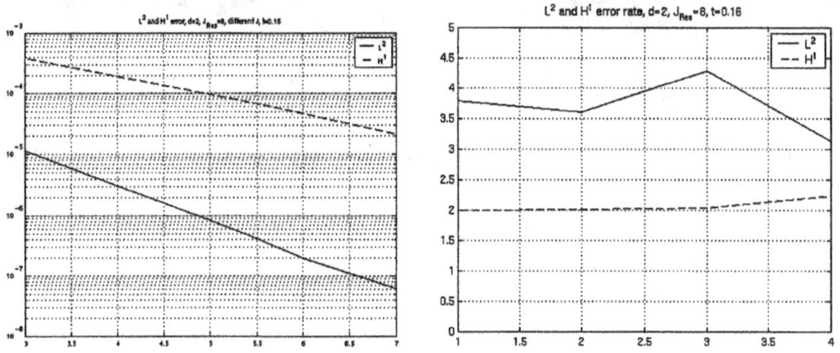

Fig. 3.22. L_2- and H^1-error over the maximal level for $t = 0.16$.

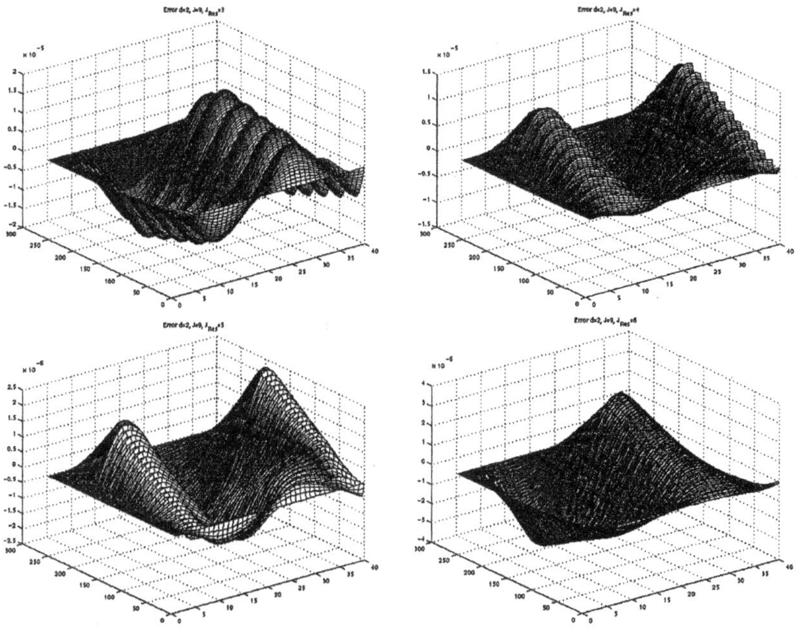

Fig. 3.23. Error for $J_{\text{Res}} = 3, 4, 5, 6$, $d = 2$.

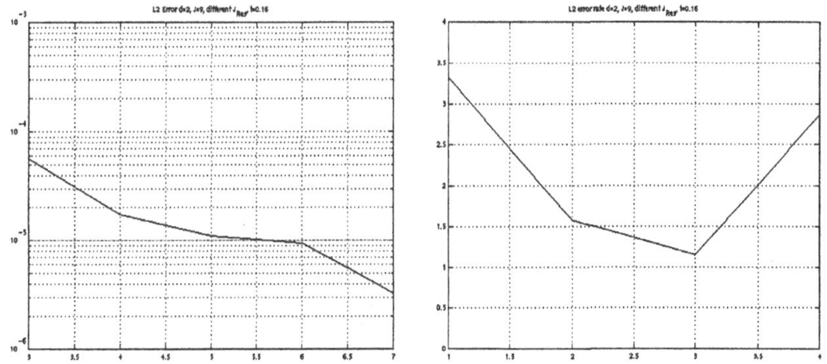

Fig. 3.24. L_2-error (left) and rate of convergence (right) w.r.t. J_{Res} at $t = 0.16$.

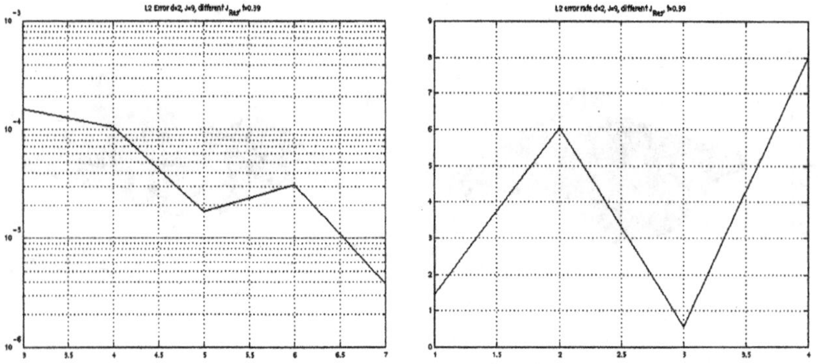

Fig. 3.25. L_2-error (left) and rate of convergence (right) w.r.t. J_{Res} at $t = 0.39$.

3.3.5 Numerical Results II: Plastic Indicators

The observations of the first experiment above have motivated the following choice for plastic indicators. Since in the elastic regime all involved data are smooth, one can expect that $u \in H^2$ locally in these areas. The above experiment shows a certain loss of smoothness when going from an elastic to a plastic area. Hence, referring to (1.26) on p. 9, we have considered the weightened sequence of wavelet coefficients

$$c_\lambda := 2^{2|\lambda|} d_\lambda \tag{3.55}$$

whose ℓ_2-norm is equivalent to the H^2-norm of the corresponding function (if the wavelets are sufficiently smooth). Hence, we investigate the size of these coefficients c_λ in order to resolve local effects.

In order to seperate effects due to the boundary conditions, we have changed the input data in such a form that we obtain plastic waves starting at the boundaries and a clearly separated plastic wave in the interior of the intervall. The test data have been choosen as

$$\begin{array}{lll} T = 0.25, & \Delta t = 0.01, & \sigma_{Y_1} = \sigma_{Y_2} = 1600, \\ A(x) \equiv 100, & \rho(x) \equiv 7.85, & E_Y = 2100000, \\ E_t = 200000, & f(t,x) := a\frac{t}{T}(x - \frac{1}{2}), & a = 5 \cdot 10^7. \end{array}$$

The arising plastic indicators are shown in Fig. 3.26 (right) whereas the displacement is visualized on the left in Fig. 3.26. Again, all hardening cases appear.

In Figs. 3.27 to 3.30, the scaled wavelet coefficients are shown for the cases $d = \tilde{d} = 2$, $d = 2$, $\tilde{d} = 8$, $d = 3$, $\tilde{d} = 5$, and $d = 4$, $\tilde{d} = 8$, respectively. Starting from the upper left picture which corresponds to $t = 0.1$ we show columnwise the evolution until $t = T = 0.25$ in the lower right corner.

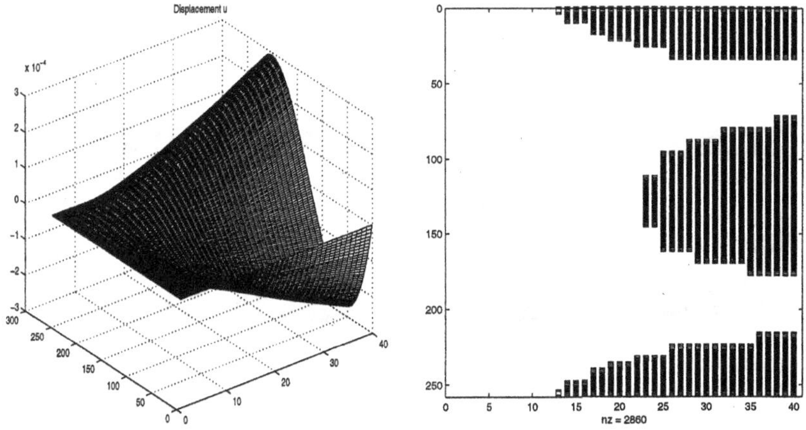

Fig. 3.26. Displacement (left) and plastic indicators (right) for the second experiment.

In the first case ($d = \tilde{d} = 2$, Fig. 3.27), the interpolatory wavelet basis and test and trial bases for the Wavelet-Galerkin method essentially coincide. This means that no extra L_2-projection has to be performed. Moreover, these functions are best localized, i.e., they have the smallest support among biorthogonal B-spline wavelets. On the other hand, the H^2-norm equivalence does not hold for piecewise linear B-spline wavelets since they are not sufficiently smooth. Still, $\|u\|_{2,\Omega}$ is bounded by the weightened sequence norm (but not vice versa).

We compare these pictures with two different kind of results, namely higher order of vanishing moments and higher order discretizations. In Fig. 3.28, we have shown the case $d = 2$, $\tilde{d} = 8$, i.e., piecewise linear test and trial functions with high order of vanishing moments. It is interesting to see that even though the support of these functions are much less local than in the first case, no 'smearing' effect appears. The wavelet coefficients still clearly reflect the regions of plasticity.

Interesting effects appear when going to higher order discretizations as shown for $d = 3$, $\tilde{d} = 5$ in Fig. 3.29 and $d = 4$, $\tilde{d} = 8$ in Fig. 3.30. In both cases, the H^2-norm equivalence holds. As we see, the scaled wavelet coefficients of both systems reflect the *interface* of the plastic region which gives rise to a much sharper and sparser description of the hardening process. This shows two things: First the loss of the optimal rate of convergence as observed in the first experiment is only due to the (small) transition regions. Second, wavelet coefficients of high order discretizations give a sharp description of plastic waves. Both results strongly indicate the potential for adaptive wavelet methods.

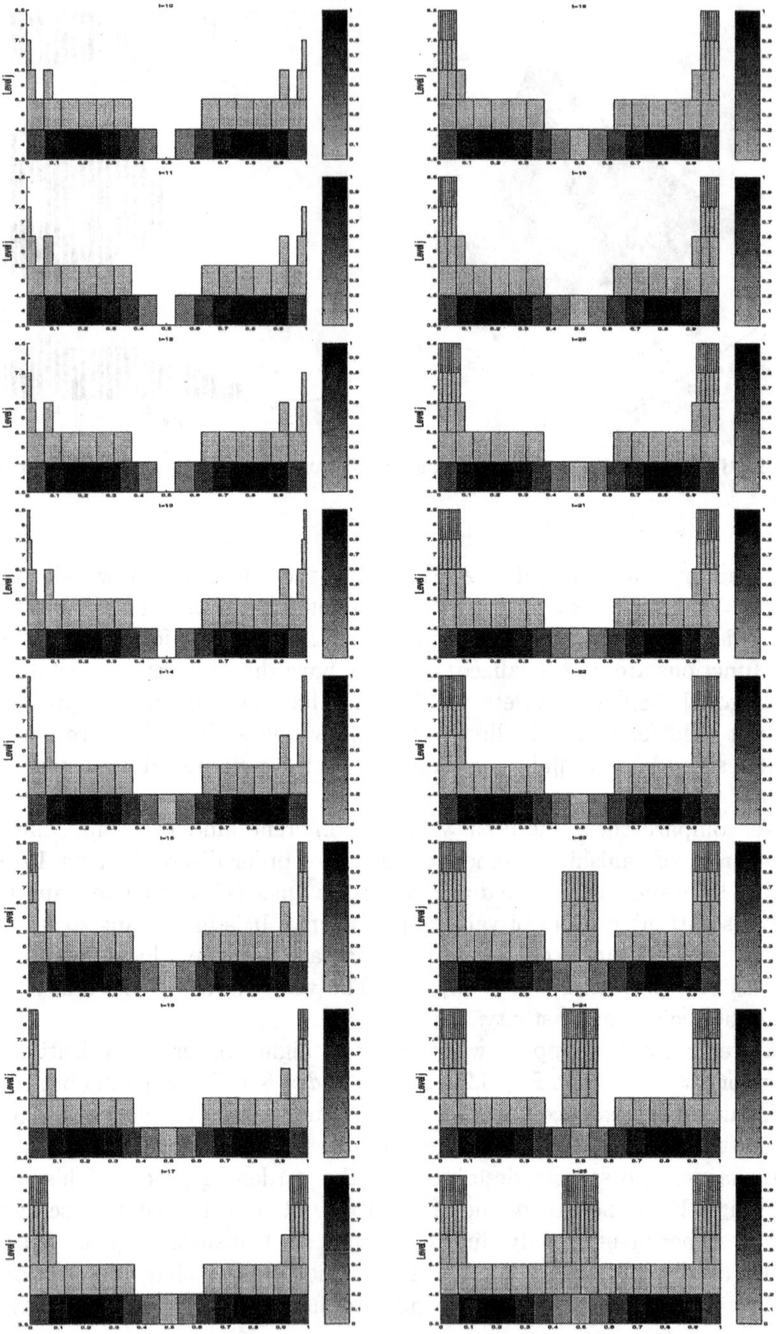

Fig. 3.27. Wavelet coefficients for $d = 2$, $\tilde{d} = 2$ and $t = 10, \ldots, 25$.

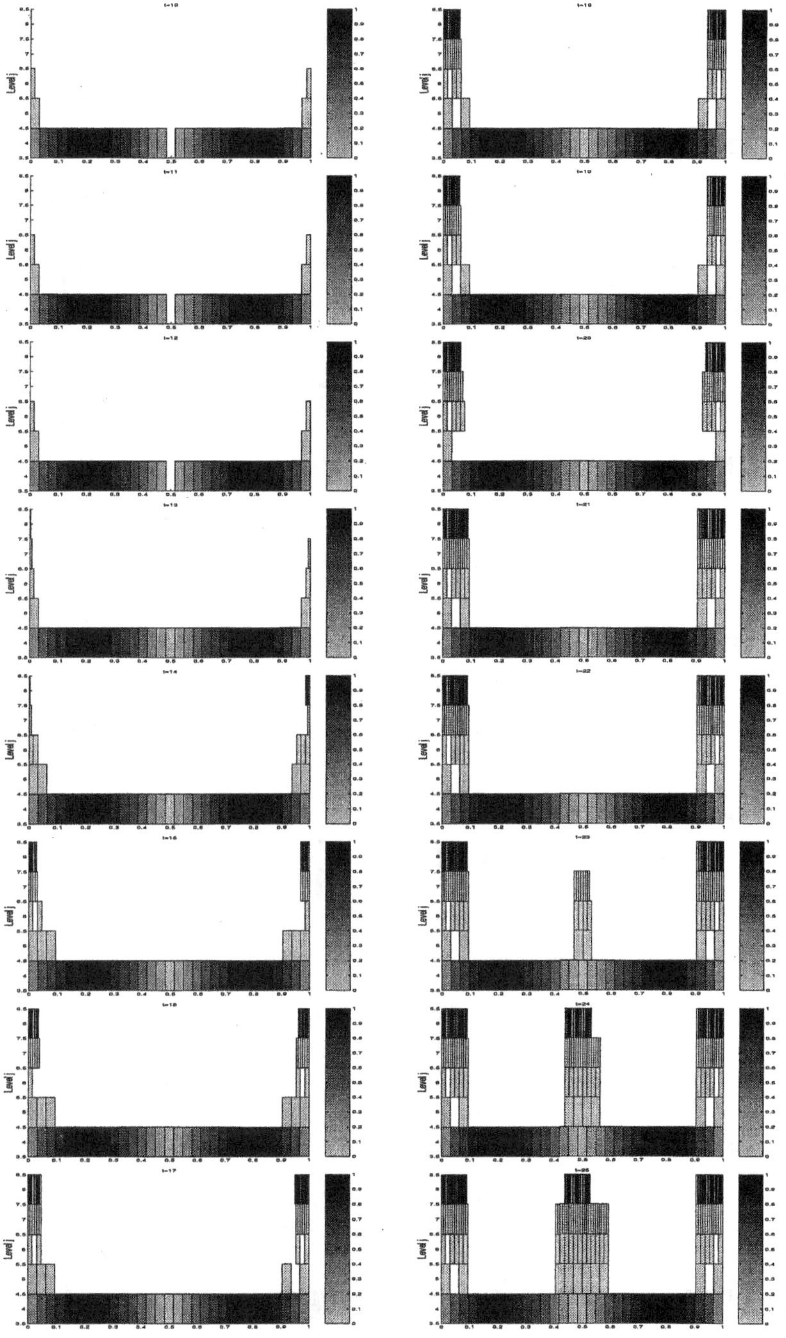

Fig. 3.28. Wavelet coefficients for $d = 2$, $\tilde{d} = 8$ and $t = 10, \ldots, 25$.

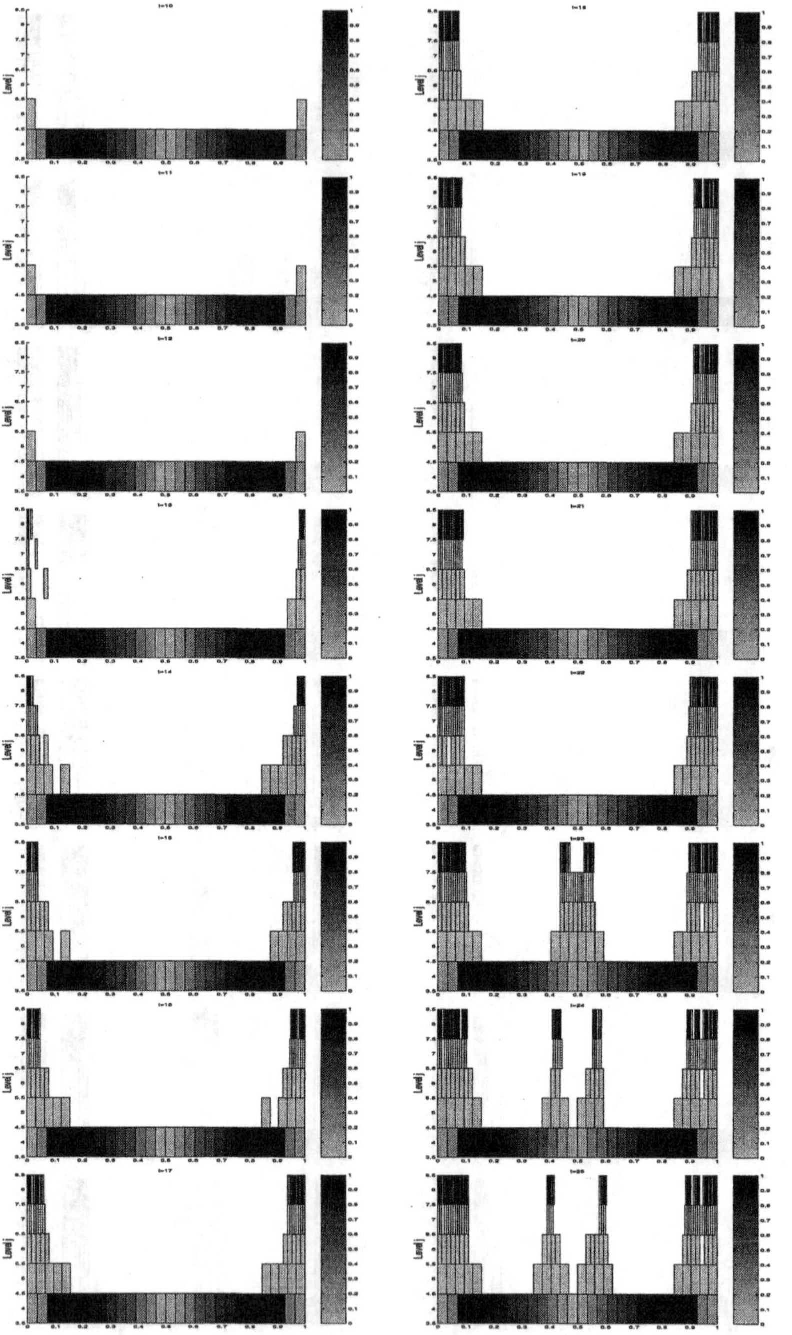

Fig. 3.29. Wavelet coefficients for $d = 3$, $\tilde{d} = 5$ and $t = 10, \ldots, 25$.

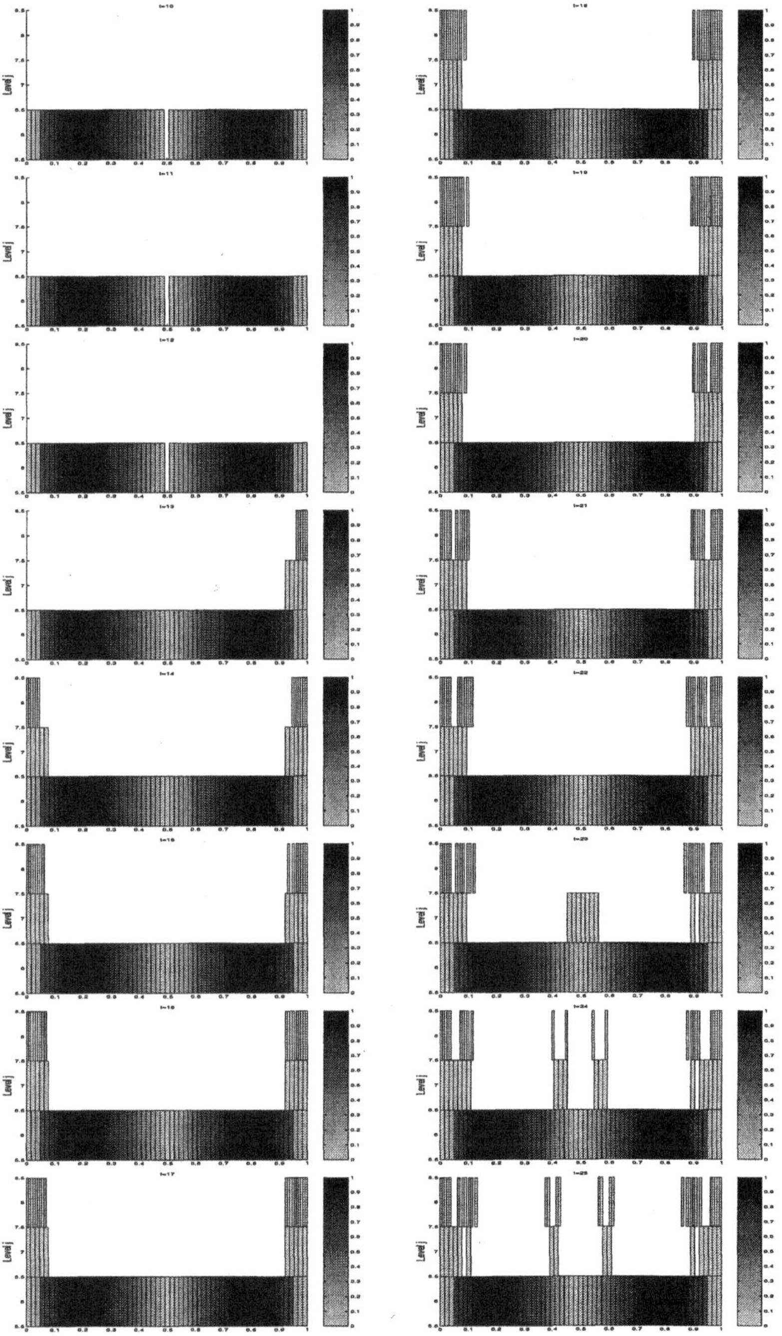

Fig. 3.30. Wavelet coefficients for $d = 4$, $\tilde{d} = 8$ and $t = 10, \dots, 25$.

References

1. Albukrek, C.M., Urban, K., Rempfer, D., Lumley, J.L. (2000): Divergence-free Wavelet Analysis of Turbulent Flow. RWTH Aachen, IGPM Preprint 198. To appear in J. Scient. Comput.
2. Andersson, L., Hall, N., Jawerth, B., Peters, G. (1994): Wavelets on Closed Subsets of the Real Line. In: Schumaker, L.L., Webb, G. (eds) Topics in the Theory and Applications of Wavelets, Academic Press, Boston, 1994, 1–61
3. Arnold, D.N., Falk, R.S., Winther R. (1997): Preconditioning in H(div) and Applications. Math. Comput., **66**, No. 219, 957–984
4. Arnold, D.N., Falk, R.S., Winther R. (2000): Multigrid in H(div) and H(**curl**). Numer. Math., **85**, No. 2, 197–217
5. Barinka, A., Barsch, T., Charton, P., Cohen, A., Dahlke, S., Dahmen, W., Urban, K. (1999): Adaptive Wavelet Schemes for Elliptic Problems — Implementation and Numerical Experiments. RWTH Aachen, IGPM Preprint 173. To appear in SIAM J. Scient. Comput.
6. Barinka, A., Barsch, T., Urban, K., Vorloeper, J. (1998): The Multilevel Library: Software Tools for Multiscale Methods and Wavelets, Version 1.0, Documentation. RWTH Aachen, IGPM Preprint 156
7. Barsch, T. (2000): Adaptive Multiskalenverfahren für elliptische partielle Differentialgleichungen — Realisierung, Umsetzung und numerische Ergebnisse (in german). PhD Thesis, RWTH Aachen, Shaker Aachen
8. Barsch, T., Kunoth, A., Urban, K. (1997): Towards Object Oriented Software Tools for Numerical Multiscale Methods for P.D.E.s using Wavelets. In: Dahmen, W., Kurdilla, A., Oswald, P. (eds) Multiscale Methods for Partial Differential Equations, Academic Press, 383–412
9. Barsch, T., Urban, K. (2000): Software Tools for Using Wavelets on the Interval for the Numerical Solution of Operator Equations. In: Hackbusch, W., Wittum, G. (eds) Concepts of Numerical Software, Kiel, 13–15
10. Beck, R., Deuflhard, P., Hiptmair, R., Hoppe, R.H.W., Wohlmuth, B. (1999): Adaptive Multilevel Methods for Edge Element Discretizations of Maxwell's Equations. Surv. Math. Ind., **8**, No. 3-4, 271–312
11. Bergh, J., Löfström, J. (1976): Interpolation Spaces, An Introduction, Springer New York
12. Berkooz, G., Elezgaray, J., Holmes, P.J. (1992): Coherent Structures in Random Media and Wavelets. Physica D, **61**, 47–58
13. Berkooz, G., Holmes, P., Lumley, J.L. (1993): The Proper Orthogonal Decomposition in the Analysis of Turbulent Flows. Ann. Rev. Fluid Mech., **25**, 539–575
14. Bertoluzza, S., Falletta, S. (2001): Wavelets on]0,1[at large scales. I.A.N. Pavia, Preprint no. 1212

15. Bertoluzza, S., Naldi, G. (1996): A Wavelet Collocation Method for the Numerical Solution of Partial Differential Equations. Appl. Comp. Harm. Anal., **1**, 1–9
16. Beylkin, G., Coifman, R., Rokhlin, V. (1991): Fast wavelet Transforms and Numerical Algorithms. Comm. Pure Appl. Math., **44**, 141–183.
17. Bossavit, A. (1998): Computational Electromagnetism. Academic Press San Diego
18. Braess, D. (1997): Finite Elements: Theory, Fast Solvers and Applications in Solid Mechanics. Cambridge University Press, Cambridge
19. Brezzi, F., Marini, L.D. (1994): A Three-field Domain Decomposition Method. Contemp. Math., **157**, 27–34
20. Canuto, C., Hussaini, M.Y., Quarteroni, A., Zang, T.A. (1988): Spectral Methods in Fluid Dynamics. Springer New York
21. Canuto, C., Tabacco, A. (2000): Absolute and Relative Cut-off in Adaptive Approximations by Wavelets. Ann. Mat. Pura Appl., **178**, 287–315
22. Canuto, C., Tabacco, A., Urban, K. (1998): Numerical Solution of Elliptic Problems by the Wavelet Element Method. In: Bock, H.G. et al. (eds) ENUMATH 97, World Scientific Singapore, 17–37
23. Canuto, C., Tabacco, A., Urban, K. (1999): The Wavelet Element Method, Part I: Construction and Analysis. Appl. Comp. Harm. Anal., **6**, 1–52
24. Canuto, C., Tabacco, A., Urban, K. (2000): The Wavelet Element Method, Part II: Realization and Additional Features in 2d and 3d. Appl. Comp. Harm. Anal., **8**, 123–165
25. Carnicer, J.M., Dahmen, W., Peña, J.M. (1996): Local Decomposition of Refinable Spaces. Appl. Comp. Harm. Anal., **3**, 127–153
26. Chiavassa, G., Liandrat, J. (1997): On the Effective Construction of Compactly Supported Wavelets Satisfying Homogeneous Boundary Conditions on the Interval. Appl. Comput. Harm. Anal., **4**, No. 1, 62–73
27. Cohen, A. (2000): Wavelet methods in Numerical Analysis. In: Ciarlet, P.G., Lions, J.L. (eds) Handbook of Numerical Analysis. Elsevier North-Holland, 417–711
28. Cohen, A., Dahmen, W., DeVore, R. (2001): Adaptive Wavelet Schemes for Elliptic Operator Equations — Convergence Rates. Math. Comput., **70**, No. 233, 27–75
29. Cohen, A., Dahmen, W., DeVore, R. (2000): Adaptive Wavelet Methods II — Beyond the Elliptic Case. RWTH Aachen, IGPM Preprint 199
30. Cohen, A., Daubechies, I., Feauveau, J.-C. (1992): Biorthogonal Bases of Compactly Supported Wavelets. Comm. Pure Appl. Math., **45**, 485–560
31. Cohen, A., Daubechies, I., Vial, P. (1993): Wavelets on the Interval and Fast Wavelet Transforms. Appl. Comp. Harm. Anal., **1**, 54–81
32. Cohen, A., Masson, R. (1999): Wavelet Methods for Second Order Elliptic Problems — Preconditioning and Adaptivity. SIAM J. Sci. Comp., **21**, No. 3, 1006–1026
33. Cohen, A., Masson, R. (2000): Wavelet Adaptive Methods for Second Order Elliptic Problems — Boundary Conditions and Domain Decomposition. Numer. Math., **86**, No. 2, 193–238
34. Cohen, A., Schlenker, J.-M. (1993): Compactly Supported Wavelet Bases with Hexagonal Symmetry. Constr. Approx., **9**, No. 2-3, 209–236

35. Comincioli, V., Naldi, G., Scapolla, T., Venini, P. (2000): A Wavelet like Galerkin Method for Numerical Solution of Variational Inequalities Arising in Elastoplasticity. To appear in Commun. Numer. Methods Eng.
36. Costabel, M. (1988): Boundary Integral Operators on Lipschitz Domains: Elementary Results. SIAM J. Math. Anal., **19**, 613–626
37. Dahlke, S. (1999): Besov Regularity for Elliptic Boundary Value Problems on Polygonal Domains. Appl. Math. Lett., **12(6)**, 31–36
38. Dahlke, S., Dahmen, W., DeVore, R.A. (1997): Nonlinear Approximation and Adaptive Techniques for Solving Elliptic Operator Equations. In: Dahmen, W., Kurdilla, A., Oswald, P. (eds) Multiscale Wavelet Methods for PDEs. Academic Press San Diego, 237–283
39. Dahlke, S., Dahmen, W., Hochmuth, R., Schneider, R. (1997): Stable Multiscale Bases and Local Error Estimation for Elliptic Problems. Appl. Numer. Math., **23**, No. 1, 21–48
40. Dahlke, S., Dahmen, W., Urban, K. (2000): Adaptive Wavelet Methods for Saddle Point Problems — Convergence Rates. RWTH Aachen, IGPM Preprint 205
41. Dahlke, S., DeVore, R.A. (1997): Besov Regularity for Elliptic Boundary Value Problems. Comm. Part. Diff. Eq., **22(1&2)**, 1–16
42. Dahlke, S., Hochmuth, R., Urban, K (2000): Adaptive Wavelet Methods for Saddle Point Problems. Math. Model. Numer. Anal., **34**, No. 5, 1003–1022
43. Dahlke, S., Hochmuth, R., Urban, K. (2000): Convergent Adaptive Wavelet Methods for the Stokes Problem. In: Dick, E., Riemslagh, E., Vierendeels, J. (eds) Multigrid Methods VI. Springer Heidelberg Berlin
44. Dahmen, W. (1993): Decomposition of Refinable Spaces and Applications to Operator Equations. Numer. Alg., **5**, 229–245
45. Dahmen, W. (1995): Multiscale Analysis, Approximation, and Interpolation Spaces. In: Chui, C.K., Schumaker, L.L. (eds) Approximation Theory VIII. World Scientific Singapore, 47–88
46. Dahmen, W. (1996): Stability of Multiscale Transformations. J. Fourier Anal. and Appl., **2**, 341–361
47. Dahmen, W., (1997): Wavelet and Multiscale Methods for Operator Equations. Acta Numerica **6**, 55–228
48. Dahmen, W. (2001): Wavelet Methods for PDEs — Some Recent Developments. In: Sloan, D., Van de Walle, S., Süli, E. (eds) Partial differential equations. Elsevier North-Holland 2001.
49. Dahmen, W., Han, B., Jia, R.-Q., Kunoth, A. (2000): Biorthogonal Multiwavelets on the Interval: Cubic Hermite Splines. Constr. Approx., **16**, 221–259
50. Dahmen, W., Kunoth, A. (1992): Multilevel Preconditioning. Numer. Math., **63**, 315–344
51. Dahmen, W., Kunoth, A. (2001): Appending Boundary Conditions by Lagrange Multipliers: General Criteria for the LBB condition. Numer. Math., **88**, 9–42
52. Dahmen, W., Kunoth, A., Urban, K. (1996): A Wavelet-Galerkin Method for the Stokes Problem. Computing, **56**, 259–302
53. Dahmen, W., Kunoth, A., Urban, K. (1997): Wavelets in Numerical Analysis and their Quantitative Properties. In: Le Mehaute, A., Rabut, C., Schumaker, L.L. (eds) Surface Fitting and Multiresolution Methods. Vanderbilt Univ. Press, Nashville, 93–130

54. Dahmen, W., Kunoth, A., Urban, K. (1999): Biorthogonal Spline-wavelets on the Interval — Stability and Moment Conditions. Appl. Comp. Harm. Anal., **6**, 132–196
55. Dahmen, W., Micchelli, C.A. (1993): Banded Matrices with Banded Inverses, II: Locally Finite Decomposition of Spline Spaces. Constr. Appr., **9**, 263–281
56. Dahmen, W., Micchelli, C.A. (1993): Using the Refinement Equation for Evaluating Integrals of Wavelets. Siam J. Numer. Anal., **30**, 507–537
57. Dahmen, W., Prößdorf, S., Schneider, R. (1994): Multiscale Methods for Pseudo-differential Equations on Smooth Manifolds. In: Chui, C.K., Montefusco, L., Puccio, L. (eds) Proceedings of the International Conference on Wavelets: Theory, Algorithms, and Applications. Academic Press San Diego, 385–424
58. Dahmen, W., Schneider, R. (1998): Wavelets with Complementary Boundary Conditions — Functions Spaces on the Cube, Results Math., **34**, 255–293
59. Dahmen, W., Schneider, R. (1999): Composite Wavelet Bases for Operator Equations. Math. Comput., **68**, 1533–1567
60. Dahmen, W., Schneider, R. (1999): Wavelets on Manifolds I: Construction and Domain Decomposition. SIAM J. Math. Anal., **31**, 184–230
61. Dahmen, W., Schneider, R., Xu, Y. (2000): Nonlinear Functionals of Wavelet Expansions — Adaptive Reconstruction and Fast Evaluation. Numer. Math., **86**, 49–101
62. Dahmen, W., Stevenson, R. (1999): Element-by-element Construction of Wavelets Satisfying Stability and Moment Conditions, SIAM J. Numer. Anal., **37**, 319–325
63. Daubechies, I. (1992): Ten Lectures on Wavelets. CBMS-NSF Regional Conference Series in Applied Mathematics **61**
64. Dautray, W., Lions, J.L. (1993): Mathematical Analysis and Numerical Methods for Science and Technology. Springer Heidelberg Berlin
65. Deslaurier, G., Debuc, S. (1989): Symmetric Interpolation Processes. Constr. Approx., **5**, 49-68
66. DeVore, R.A. (1998): Nonlinear Approximation, Acta Numerica, **7**, 51–150
67. DeVore, R.A., Lucier, B. (1990): High Order Regularity for Conservation Laws, Indiana Math. J., **39**, 413–430
68. DeVore, R.A., Lucier, B. (1996): On the Size and Smoothness of Solutions to Nonlinear Hyperbolic Conservation Laws. SIAM J. Math. Anal., **27**, 684–707
69. DeVore, R.A., Popov, V. (1998): Interpolation Spaces and Nonlinear Approximation. In: Cwikel, M., Peetre, J., Sagher, Y., Wallin, H. (eds) Function Spaces and Approximation. Springer Lecture Notes in Math., 191–205
70. Donoho, D. (1992): Interpolating Wavelet Transform. Department of Statistics, Stanford University, Preprint
71. Farge, M. (1992): Wavelet Transforms and their Applications to Turbulence. Ann. Rev. Flu. Mech., **24**, 395–457
72. Farge, M. (1992): The Continuous Wavelet Transform of Two-dimensional Turbulent Flows. In: Ruskai, M.B. et al. (eds) Wavelets and their Applications. Jones and Bartlett Boston, 275–302
73. Farge, M., Schneider, K., Kevlahan, N. (1999): Non-Gaussianity and Coherent Vortex Simulation for Two-dimensional Turbulence using an Adaptive Orthogonal Wavelet Basis. Physics of Fluids, **11**, No. 8, 2187–2201
74. Farouki, R.T., Rajan, V.T. (1987): On the numerical condition of polynomials in Bernstein form. Computer Aided Geometric Design, **4**, 191–216

75. Fletcher, C.A.J. (1988): Computational Techniques for Fluid Dynamics, Volume II. Springer Berlin
76. Frazier, M., Jawerth, B. (1985): Decomposition of Besov Spaces, Indiana Math. J., **34**, 777–799
77. Girault, V., Raviart, P.-A. (1986): Finite Element Methods for Navier-Stokes-Equations. 2nd edition, Springer Berlin
78. Gordon, W., Hall, C. (1973): Construction of Curvlinear Co-ordinate Systems for Nonlinear Problems in Fluid Dynamics. Comput. Meth. Appl. Mech. Eng., **7**, 461–477
79. Gordon, W., Hall, C. (1973): Transfinite Element Methods: Blending-function Interpolation over Arbitrary Curved Element Domains. Numer. Math., **21**, 109–129
80. Griebel, M., Dornseifer, T., Neunhoeffer, T. (1995) Numerische Simulation in der Strömungsmechanik (in german). Vieweg Braunschweig
81. Griebel, M., Koster, F. (2000): Adaptive Wavelet Solvers for the Unsteady Incompressible Navier-Stokes Equations. In: Malek, J., Rokyta, M. (eds) Advances in Mathematical Fluid Mechanics, Springer Heidelberg, 67–118
82. Grivet Talocia, S., Tabacco, A. (2000): Wavelets on the Interval with Optimal Localization. Math. Models Meth. Appl. Sci., **10**, 441–462
83. Hackl, K. (1996): A Wavelet Based Elastoplasticity Beam Model. Z. Angew. Math. Mech., **76** suppl. 1, 175–178
84. Han, W., Reddy, B.D. (1995): On the Finite Element Method for Mixed Variational Inequalities Arising in Elastoplasticity. SIAM J. Numer. Anal., **32**, No. 6, 1778-1807
85. Hardin, D.P., Marasovich, J.A. (1999): Biorthogonal Multiwavelets on $[-1,1]$. Appl. Comp. Harm. Anal., **7**, No. 1, 34–53
86. Hiptmair, R. (1996): Multilevel Preconditioning for Mixed Problems in Three Dimensions. PhD Thesis, Univ. of Augsburg
87. Holmes, P.J., Berkooz, G., Lumley, J.L. (1996): Turbulence, Coherent Structures, Dynamical Systems and Symmetry. Cambridge University Press
88. Jaffard, S. (1992): Wavelet Methods for Fast Resolution of Elliptic Problems. SIAM J. Numer. Anal., **29**, 965–986
89. Jia, R.-Q., Micchelli, C.A. (1991): Using the Refinement Equations for the Construction of Pre-wavelets II: Powers of Two. In: Laurent, P.J., Le Méhauté, A., Schumaker, L.L. (eds) Curves and Surfaces. Academic Press San Diego
90. Kolmogorov, A. (1942): The Equations of Turbulent Motion in an Incompressible Fluid. Isv. Sci. USSR Phys., **6**, 56–58
91. Koster, F., Schneider, K., Griebel, M., Farge, M. (2001): Adaptive Wavelet Methods for the Navier-Stokes Equations. To appear in: Hirschel, E.H. (ed) Notes on Numerical Fluid Mechanics. Vieweg Braunschweig
92. Kunoth, A. (1995): Multilevel Preconditioning — Appending Boundary Conditions by Lagrange Multipliers. Adv. Comput. Math., **4**, No. 1,2, 145–170
93. Lakey, J.D., Massopust, R., Pereyra, M.C. (1998): Divergence-free Multiwavelets. In: Chui, C.K., Schumaker, L.L. (eds) Approximation Theory IX, Vol. 2. World Scientific Singapore, 161-168
94. Lakey, J.D., Pereyra, M.C. (1999): Multiwavelets on the Interval and Divergence-free Wavelets. In: Proc. of the SPIE **3813** 'Wavelet Applications in Signal and Image Processing VII', Denver, 162–173

95. Lakey, J.D., Pereyra, M.C. (2000): Divergence-free Multiwavelets on Rectangular Domains. In: Tian-Xiao, H. (ed) Wavelet analysis and multiresolution methods. Dekker Lect. Notes Pure Appl. Math. **212**, 203–240
96. Lakey, J.D., Pereyra, M.C. (2001): On the Non-existence of Certain Divergence-free Multiwavelets. To appear in: Debnath, L. (ed) Wavelet Transforms an Time-Frequency Signal Analysis
97. Launder, B., Spalding, D. (1972): Mathematical Models of Turbulence. Academic Press San Diego
98. Lemarié-Rieusset, P.G. (1992): Analyses Multi-résolutions non Orthogonales, Commutation entre Projecteurs et Derivation et Ondelettes Vecteurs à Divergence Nulle (in french). Rev. Mat. Iberoamericana **8**, 221–236
99. Lemarié-Rieusset, P.G. (1994): Un théorème d'Inexistence pour les Ondelettes Vecteurs à Divergence Nulle (in french). C.R. Acad. Sci. Paris Sér. I Math., **319**, 811–813
100. Lewalle, J. (1993): Wavelet Transforms of the Navier-Stokes Equations and the Generalized Dimensions of Turbulence. Appl. Sci. Res., **51**, No. 1-2, 109–113
101. Lewalle, J. (1993): Energy Dissipation in the Wavelet-transformed Navier-Stokes Equations. Phys. Fluids, **A 5**, No. 6, 1512–1513
102. Lumley, J.L. (1967): The Structure of Inhomogeneous Turbulence. In: Yaglom, A.M., Tatarski, V.I. (eds) Atmospheric turbulence and wave propagation. Nauka Moscow
103. Mallat, S.G. (1998): A Theory for Multiresolution Signal Decomposition: The wavelet Representation. IEEE Trans. Pattern Anal. Machine Intel., **11**(7), 674–693
104. Mazya, V.G. (1991): Boundary Integral Equations. In: Mazya, V.G., Nikol'skii, S.M. (eds) Encyclopaedia of Math. Sciences **27**, Analysis IV. Springer Berlin Heidelberg
105. Meneveau, C. (1991): Analysis of Turbulence in the Orthonormal Wavelet Representation. J. Fluid Mech., **232**, 469–520
106. Naldi, G., Urban, K., Venini, P. (2000): A Wavelet-Galerkin Method for Elasticity Problems. Math. Modell. Sci. Comput., **10**
107. Naldi, G., Venini, P. (1997): Wavelet Analysis of Structures: Statics, Dynamics and Damage Identification. Meccanica, **32**, 223–230
108. Meyer, Y. (1992): Wavelets and Operators. Cambridge Studies in Advanced Mathematics **37**
109. Owen, D.R.J., Hinton, E. (1980): Finite Elements in Plasticity Theory and Practice. Pineridge Press, Swansea
110. Riemenschneider, S.D., Shen, Z. (1992): Wavelets and Pre-wavelets in Low Dimensions. J. Approx. Theory, **71**, No. 1, 18–38
111. Roberts, J.E., Thomas, J.-M. (1991): Mixed and Hybrid Methods, In: Ciarlet, P.G., Lions, J.L. (eds) Handbook of Numerical Analysis, Vol. II. Elsevier North-Holland, 523–640
112. Schatz, A., Thomée, V., Wendland, W. (1990): Mathematical Theory of Finite and Boundary Element Methods. Birkhäuser Basel
113. Simo, J.C., Hughes, T.J.R. (1998): Computational Inelasticity. Springer New York
114. Smagorinsky, J. (1963): General Circulation Model of the Atmosphere. Mon. Whether Rev. **91**, 99–164
115. Tennekes, H., Lumley, J.L. (1972): A First Course in Turbulence. MIT Press Cambridge, MA

116. Urban, K. (1995): On Divergence-free Wavelets. Adv. Comput. Math., **4**, 51–82
117. Urban, K. (1995): Multiskalenverfahren für das Stokes-Problem und angepaßte Wavelet-Basen (in german). PhD Thesis, Verlag der Augustinus-Buchhandlung, Aachen
118. Urban, K. (2001): Wavelet Bases in $\boldsymbol{H}(div)$ and $\boldsymbol{H}(\boldsymbol{curl})$. Math. Comput., **70**, 739–766
119. Yserentant, H. (1993): Old and New Convergence Proofs for Multigrid Methods. Acta Numerica, 285–326
120. Zienkiewicz, O.C., Taylor, R.L. (1991): Finite Element Method: Solid and Fluid Mechanics Dynamics and Non-Linearity. McGraw-Hill

List of Figures

List of Tables

List of Symbols

General:

$\boldsymbol{x}^T$	transpose of a vector or matrix
$\boldsymbol{A}^{\ddagger}$	counter-transpose of a matrix $\boldsymbol{A}$
$\lfloor x \rfloor$ ($\lceil x \rceil$)	largest (smallest) integer less (greater) than or equal to x
$\boldsymbol{I}_X$	identity matrix w.r.t. the index set X
$\delta_{i,j}$	Kronecker symbol
$\boldsymbol{\delta}_i$	i-th canonical vector of unity $(\delta_{1,i}, \ldots, \delta_{n,i})^T$
$X^{\perp}$	orthogonal complement
∂_i	short for partial derivative $\frac{\partial}{\partial x_i}$
$GL(X, Y)$	space of bounded linear mappings from X to Y
DF, JF	Jacobi matrix of a mapping F
$X \hookrightarrow Y$	X is continuously embedded into Y
clos	closure
span	linear span
diam	diameter
$a \lesssim b$	$a \leq \mathrm{const}\, b$
$a \sim b$	$a \lesssim b$ and $b \lesssim a$
$\mathcal{F}g$	Fourier transform of g
F^*	adjoint operator
B_r^d	r-th Bernstein polynomial of degree d

Domains:

Ω	bounded domain in $\mathbb{R}^n$
$\Gamma = \partial\Omega$	boundary of Ω
Γ_{Dir}	Dirichlet part of Γ
Γ_{Neu}	Neumann part of Γ
$\mathcal{P}$	partition $\{\Omega_i : i = 1, \ldots, N\}$ of Ω
$\hat{\Omega}$	reference domain
σ	p-face of Ω_i

Differential operators:

curl	curl operator
curl	= rot in 2D
div	divergence operator

Function spaces:

$L_2(\Omega)$	square integrable functions on Ω
$L_2(0,1)$	square integrable functions on $(0,1)$
$L_2^{\mathrm{loc}}(\Omega)$	locally square integrable functions on Ω
$\boldsymbol{L}_2(\Omega)$	square integrable vector fields on Ω
$H^s(\Omega)$	Sobolev space of order $s \in \mathbb{R}$ on Ω
$W_p^m(\Omega)$	Sobolev space of order $s \in \mathbb{R}$ on Ω with measure p
$B_p^s(L_q(\Omega))$	Besov space of order $s \in \mathbb{R}$ with measurement p, q
$H_b^s(\Omega; \mathcal{P})$	broken Sobolev spaces w.r.t. to a partition $\mathcal{P}$
$\boldsymbol{H}(\mathrm{div}; \Omega)$	$= \{\boldsymbol{f} \in \boldsymbol{L}_2(\Omega) : \mathrm{div}\, \boldsymbol{f} \in L_2(\Omega)\}$
$\boldsymbol{H}(\mathbf{curl}; \Omega)$	$= \{\boldsymbol{f} \in \boldsymbol{L}_2(\Omega) : \mathbf{curl}\, \boldsymbol{f} \in L_2(\Omega)\}$
$\boldsymbol{V}(\mathbf{curl}; \Omega)$	**curl**-free vector fields in $\boldsymbol{H}(\mathbf{curl}; \Omega)$
$\boldsymbol{V}(\mathrm{div}; \Omega)$	divergence-free vector fields in $\boldsymbol{H}(\mathrm{div}; \Omega)$
$\boldsymbol{H}_0(\mathbf{curl}; \Omega)$	$\boldsymbol{H}(\mathbf{curl}; \Omega)$-vector fields with vanishing trace
$\boldsymbol{H}_0(\mathrm{div}; \Omega)$	$\boldsymbol{H}(\mathrm{div}; \Omega)$-vector fields with vanishing trace
$\mathcal{P}_r(\Omega)$	space of algebraic polynomials on Ω of degree $r \in \mathbb{N}$
ℓ_p	space of p-sumable sequences

Norms and inner products:

X'	dual space
$\|\cdot\|_{0,\Omega}$	norm on $L_2(\Omega)$
$\|\cdot\|_{s,\Omega}$	norm on $H^s(\Omega)$
$(\cdot,\cdot)_{0,\Omega}$	inner product in $L_2(\Omega)$
(v, Φ)	vector of inner products of v with all elements in Φ
$\langle\cdot,\cdot\rangle$	dual form
$\langle\cdot\rangle$	averaging process

Wavelets:

λ	wavelet index $\lambda = (j, k)$
j	level index
$f_{[j,k]}$	scaled translate $2^{j/2} f(2^j x - k)$
$\lvert\lambda\rvert$	level of a scaling function/wavelet, $\lambda = (j,k)$, $\lvert\lambda\rvert = j$
ξ	1D scaling function
η	1D wavelet
${}_d\xi$	cardinal B-spline of order $d \in \mathbb{N}$
${}_{d,\tilde{d}}\xi$	dual function of order $\tilde{d}$ to B-spline of order d
${}_{d,\tilde{d}}\eta$, ${}_{d,\tilde{d}}\tilde{\eta}$	biorthogonal B-spline wavelets of order d, $\tilde{d}$
Ξ_j	scaling function basis in 1D
Υ_j	wavelet basis in 1D

$\Xi_j^{\mathbb{R}}$, $\Upsilon_j^{\mathbb{R}}$	wavelet basis in 1D on the real line $\mathbb{R}$
φ	scaling function in nD
ψ	wavelet in nD
$\tilde{\varphi}$, $\tilde{\psi}$	dual (biorthogonal) functions
Φ_j	scaling function system in nD
Ψ_j	wavelet system in nD
I, J	scaling function/wavelet index sets in 1D
$\mathcal{I}$, $\mathcal{J}$	scaling function/wavelet index sets in nD
I_j^L, I_j^R, I_j^0	left, right and interior indices on $(0,1)$
E^*	$\{0,1\}^n \setminus \{0\}$
$\hat{\varphi}$, $\hat{\psi}$	scaling function/wavelet on the reference domain $\hat{\Omega}$
$\hat{\Phi}_j$, $\hat{\Psi}$	scaling function/wavelet system on the reference domain $\hat{\Omega}$
Ψ^{curl}	wavelets in $\boldsymbol{H}(\mathbf{curl};\Omega)$
Ψ^{div}	wavelets in $\boldsymbol{H}(\mathrm{div};\Omega)$
Ψ^{cf}	**curl**-free wavelets
Ψ^{df}	divergence-free wavelets
Ψ^{cc}	**curl**-free complement wavelets
Ψ^{dc}	divergence-free complement wavelet
$S(\Phi_j)$	$\mathrm{clos}_{L_2}\mathrm{span}\,\Phi_j$
S_j	multiresolution spaces
$\mathcal{S}$	$\{S_j\}_j$ multiresolution sequence
V_j^{β}	1D multiresolution space with boundary conditions
S_j^{b}	nD multiresolution space with boundary conditions
W_j	wavelet (detail) spaces
P_j	projectors onto S_j
Q_j	projector $P_j - P_{j-1}$ onto W_j
$\boldsymbol{M}_{j,0}$	refinement matrix
$\boldsymbol{M}_{j,1}$	wavelet refinement matrix
$\boldsymbol{M}_j$	$= (\boldsymbol{M}_{j,0}\ \boldsymbol{M}_{j,1})$ multiscale transformation matrix
$\mathcal{K}_j^i$	scaling function grid points on Ω_i
$\mathcal{H}_j^i$	wavelet grid points on Ω_i
$\mathcal{K}_j$, $\mathcal{H}_j$	scaling function/wavelet grid points on Ω
$\mathsf{d}^{(i)}$	discrete differentiation operator
grad	discrete gradient operator
div	discrete divergence operator

Fluid mechanics:

$\boldsymbol{u}$	velocity field
$\boldsymbol{u}_t$	temporal derivative of the velocity field
p	pressure
Re	Reynolds number
k	turbulent kinetic energy
ν	viscosity

Elastoplasticity:

$\boldsymbol{u}$, u	displacement
$\nabla \boldsymbol{u}$	deformation gradient
E	strain tensor
ε	strain tensor formed by symmetric derivatives
ε^e	elastic strain part
ε^p	plastic strain part
ν, μ	Lamé constants
σ	stress tensor
σ_{Y_1}	yielding stress in tension
σ_{Y_2}	yielding stress in compression
E_e	elastic modulus
E_t	tangent modulus
E_p	plastic modulus
ζ	Poisson number
φ	plasticity functions
H	hardening matrix
H_{iso}	isotropic hardening matrix
H_{kin}	kinetic hardening matrix
E_Y	Young modulus
A	cross section
ρ	mass density
f	axial force

Electromagnetism:

ρ	charge density
σ	electric conductivity
$\boldsymbol{J}_s$	suppressed current density
$\boldsymbol{E}$	electric field
$\boldsymbol{D}$	electric displacement
$\boldsymbol{H}$	magnetic field
$\boldsymbol{B}$	magnetic flux density
μ	permivity
ε	electric permeability
$\boldsymbol{cA}$	Maxwell operator

Index

Editorial Policy

§1. Volumes in the following four categories will be published in LNCSE:

i) Research monographs
ii) Lecture and seminar notes
iii) Conference proceedings
iv) Textbooks

Those considering a book which might be suitable for the series are strongly advised to contact the publisher or the series editors at an early stage.

§2. Categories i) and ii). These categories will be emphasized by Lecture Notes in Computational Science and Engineering. **Submissions by interdisciplinary teams of authors are encouraged.** The goal is to report new developments - quickly, informally, and in a way that will make them accessible to non-specialists. In the evaluation of submissions timeliness of the work is an important criterion. Texts should be well-rounded, well-written and reasonably self-contained. In most cases the work will contain results of others as well as those of the author(s). In each case the author(s) should provide sufficient motivation, examples, and applications. In this respect, Ph.D. theses will usually be deemed unsuitable for the Lecture Notes series. Proposals for volumes in these categories should be submitted either to one of the series editors or to Springer-Verlag, Heidelberg, and will be refereed. A provisional judgment on the acceptability of a project can be based on partial information about the work: a detailed outline describing the contents of each chapter, the estimated length, a bibliography, and one or two sample chapters - or a first draft. A final decision whether to accept will rest on an evaluation of the completed work which should include

- at least 100 pages of text;
- a table of contents;
- an informative introduction perhaps with some historical remarks which should be accessible to readers unfamiliar with the topic treated;
- a subject index.

§3. Category iii). Conference proceedings will be considered for publication provided that they are both of exceptional interest and devoted to a single topic. One (or more) expert participants will act as the scientific editor(s) of the volume. They select the papers which are suitable for inclusion and have them individually refereed as for a journal. Papers not closely related to the central topic are to be excluded. Organizers should contact Lecture Notes in Computational Science and Engineering at the planning stage.

In exceptional cases some other multi-author-volumes may be considered in this category.

§4. Category iv) Textbooks on topics in the field of computational science and engineering will be considered. They should be written for courses in CSE education. Both graduate and undergraduate level are appropriate. Multidisciplinary topics are especially welcome.

§5. Format. Only works in English are considered. They should be submitted in camera-ready form according to Springer-Verlag's specifications. Electronic material can be included if appropriate. Please contact the publisher. Technical instructions and/or TEX macros are available via http://www.springer.de/author/tex/help-tex.html; the name of the macro package is "LNCSE - LaTEX2e class for Lecture Notes in Computational Science and Engineering". The macros can also be sent on request.

General Remarks

Lecture Notes are printed by photo-offset from the master-copy delivered in camera-ready form by the authors. For this purpose Springer-Verlag provides technical instructions for the preparation of manuscripts. See also *Editorial Policy.*

Careful preparation of manuscripts will help keep production time short and ensure a satisfactory appearance of the finished book. The actual production of a Lecture Notes volume normally takes approximately 12 weeks.

The following terms and conditions hold:

Categories i), ii), and iii):
Authors receive 50 free copies of their book. No royalty is paid. Commitment to publish is made by letter of intent rather than by signing a formal contract. Springer-Verlag secures the copyright for each volume.

For conference proceedings, editors receive a total of 50 free copies of their volume for distribution to the contributing authors.

Category iv):
Regarding free copies and royalties, the standard terms for Springer mathematics monographs and textbooks hold. Please write to Peters@springer.de for details. The standard contracts are used for publishing agreements.

All categories:
Authors are entitled to purchase further copies of their book and other Springer mathematics books for their personal use, at a discount of 33,3 % directly from Springer-Verlag.

Addresses:

Professor M. Griebel
Institut für Angewandte Mathematik
der Universität Bonn
Wegelerstr. 6
D-53115 Bonn, Germany
e-mail: griebel@iam.uni-bonn.de

Professor D. E. Keyes
Computer Science Department
Old Dominion University
Norfolk, VA 23529–0162, USA
e-mail: keyes@cs.odu.edu

Professor R. M. Nieminen
Laboratory of Physics
Helsinki University of Technology
02150 Espoo, Finland
e-mail: rni@fyslab.hut.fi

Professor D. Roose
Department of Computer Science
Katholieke Universiteit Leuven
Celestijnenlaan 200A
3001 Leuven-Heverlee, Belgium
e-mail: dirk.roose@cs.kuleuven.ac.be

Professor T. Schlick
Department of Chemistry and
Courant Institute of Mathematical
Sciences
New York University
and Howard Hughes Medical Institute
251 Mercer Street, Rm 509
New York, NY 10012-1548, USA
e-mail: schlick@nyu.edu

Springer-Verlag, Mathematics Editorial
Tiergartenstrasse 17
D-69121 Heidelberg, Germany
Tel.: *49 (6221) 487-185
e-mail: peters@springer.de
http://www.springer.de/math/
peters.html

Lecture Notes in Computational Science and Engineering

Vol. 1 D. Funaro, *Spectral Elements for Transport-Dominated Equations.* 1997. X, 211 pp. Softcover. ISBN 3-540-62649-2

Vol. 2 H. P. Langtangen, *Computational Partial Differential Equations.* Numerical Methods and Diffpack Programming. 1999. XXIII, 682 pp. Hardcover. ISBN 3-540-65274-4

Vol. 3 W. Hackbusch, G. Wittum (eds.), *Multigrid Methods V.* Proceedings of the Fifth European Multigrid Conference held in Stuttgart, Germany, October 1-4, 1996. 1998. VIII, 334 pp. Softcover. ISBN 3-540-63133-X

Vol. 4 P. Deuflhard, J. Hermans, B. Leimkuhler, A. E. Mark, S. Reich, R. D. Skeel (eds.), *Computational Molecular Dynamics: Challenges, Methods, Ideas.* Proceedings of the 2nd International Symposium on Algorithms for Macromolecular Modelling, Berlin, May 21-24, 1997. 1998. XI, 489 pp. Softcover. ISBN 3-540-63242-5

Vol. 5 D. Kröner, M. Ohlberger, C. Rohde (eds.), *An Introduction to Recent Developments in Theory and Numerics for Conservation Laws.* Proceedings of the International School on Theory and Numerics for Conservation Laws, Freiburg / Littenweiler, October 20-24, 1997. 1998. VII, 285 pp. Softcover. ISBN 3-540-65081-4

Vol. 6 S. Turek, *Efficient Solvers for Incompressible Flow Problems.* An Algorithmic and Computational Approach. 1999. XVII, 352 pp, with CD-ROM. Hardcover. ISBN 3-540-65433-X

Vol. 7 R. von Schwerin, *Multi Body System SIMulation.* Numerical Methods, Algorithms, and Software. 1999. XX, 338 pp. Softcover. ISBN 3-540-65662-6

Vol. 8 H.-J. Bungartz, F. Durst, C. Zenger (eds.), *High Performance Scientific and Engineering Computing.* Proceedings of the International FORTWIHR Conference on HPSEC, Munich, March 16-18, 1998. 1999. X, 471 pp. Softcover. 3-540-65730-4

Vol. 9 T. J. Barth, H. Deconinck (eds.), *High-Order Methods for Computational Physics.* 1999. VII, 582 pp. Hardcover. 3-540-65893-9

Vol. 10 H. P. Langtangen, A. M. Bruaset, E. Quak (eds.), *Advances in Software Tools for Scientific Computing.* 2000. X, 357 pp. Softcover. 3-540-66557-9

Vol. 11 B. Cockburn, G. E. Karniadakis, C.-W. Shu (eds.), *Discontinuous Galerkin Methods.* Theory, Computation and Applications. 2000. XI, 470 pp. Hardcover. 3-540-66787-3

Vol. 12 U. van Rienen, *Numerical Methods in Computational Electrodynamics.* Linear Systems in Practical Applications. 2000. XIII, 375 pp. Softcover. 3-540-67629-5

Vol. 13 B. Engquist, L. Johnsson, M. Hammill, F. Short (eds.), *Simulation and Visualization on the Grid.* Parallelldatorcentrum Seventh Annual Conference, Stockholm, December 1999, Proceedings. 2000. XIII, 301 pp. Softcover. 3-540-67264-8

Vol. 14 E. Dick, K. Riemslagh, J. Vierendeels (eds.), *Multigrid Methods VI.* Proceedings of the Sixth European Multigrid Conference Held in Gent, Belgium, September 27-30, 1999. 2000. IX, 293 pp. Softcover. 3-540-67157-9

Vol. 15 A. Frommer, T. Lippert, B. Medeke, K. Schilling (eds.), *Numerical Challenges in Lattice Quantum Chromodynamics.* Joint Interdisciplinary Workshop of John von Neumann Institute for Computing, Jülich and Institute of Applied Computer Science, Wuppertal University, August 1999. 2000. VIII, 184 pp. Softcover. 3-540-67732-1

Vol. 16 J. Lang, *Adaptive Multilevel Solution of Nonlinear Parabolic PDE Systems.* Theory, Algorithm, and Applications. 2001. XII, 157 pp. Softcover. 3-540-67900-6

Vol. 17 B. I. Wohlmuth, *Discretization Methods and Iterative Solvers Based on Domain Decomposition.* 2001. X, 197 pp. Softcover. 3-540-41083-X

Vol. 18 U. van Rienen, M. Günther, D. Hecht (eds.), *Scientific Computing in Electrical Engineering.* Proceedings of the 3rd International Workshop, August 20-23, 2000, Warnemünde, Germany. 2001. XII, 428 pp. Softcover. 3-540-42173-4

Vol. 19 I. Babuška, P. G. Ciarlet, T. Miyoshi (eds.), *Mathematical Modeling and Numerical Simulation in Continuum Mechanics.* Proceedings of the International Symposium on Mathematical Modeling and Numerical Simulation in Continuum Mechanics, September 29 - October 3, 2000, Yamaguchi, Japan. 2002. VIII, 301 pp. Softcover. 3-540-42399-0

Vol. 20 T. J. Barth, T. Chan, R. Haimes (eds.), *Multiscale and Multiresolution Methods.* Theory and Applications. 2002. X, 389 pp. Softcover. 3-540-42420-2

Vol. 21 M. Breuer, F. Durst, C. Zenger (eds.), *High Performance Scientific and Engineering Computing.* Proceedings of the 3rd International FORTWIHR Conference on HPSEC, Erlangen, March 12-14, 2001. 2002. XIII, 408 pp. Softcover. 3-540-42946-8

Vol. 22 K. Urban, *Wavelets in Numerical Simulation.* Problem Adapted Construction and Applications. 2002. XV, 181 pp. Softcover. 3-540-43055-5

Printing: Strauss GmbH, Mörlenbach
Binding: Schäffer, Grünstadt